SpringerBriefs in Philosophy

SpringerBriefs present concise summaries of cutting-edge research and practical applications across a wide spectrum of fields. Featuring compact volumes of 50 to 125 pages, the series covers a range of content from professional to academic. Typical topics might include:

- A timely report of state-of-the art analytical techniques
- A bridge between new research results, as published in journal articles, and a contextual literature review
- A snapshot of a hot or emerging topic
- An in-depth case study or clinical example
- A presentation of core concepts that students must understand in order to make independent contributions

SpringerBriefs in Philosophy cover a broad range of philosophical fields including: Philosophy of Science, Logic, Non-Western Thinking and Western Philosophy. We also consider biographies, full or partial, of key thinkers and pioneers.

SpringerBriefs are characterized by fast, global electronic dissemination, standard publishing contracts, standardized manuscript preparation and formatting guidelines, and expedited production schedules. Both solicited and unsolicited manuscripts are considered for publication in the SpringerBriefs in Philosophy series. Potential authors are warmly invited to complete and submit the Briefs Author Proposal form. All projects will be submitted to editorial review by external advisors.

SpringerBriefs are characterized by expedited production schedules with the aim for publication 8 to 12 weeks after acceptance and fast, global electronic dissemination through our online platform SpringerLink. The standard concise author contracts guarantee that

- an individual ISBN is assigned to each manuscript
- each manuscript is copyrighted in the name of the author
- the author retains the right to post the pre-publication version on his/her website or that of his/her institution.

Stephanie Siewert · Katharina Kieslich ·
Matthias Braun · Peter Dabrock

Synthetic Biology and the Question of Public Participation

Governance and Ethics in Dealing with Emerging Technologies

 Springer

Stephanie Siewert
In The Research Network MaxSynBio at
the Department for Systematic Theology II
(Ethics)
The University of Erlangen–Nuremberg
Berlin, Germany

Matthias Braun
Chair of Social Ethics and Ethics
of Technology
University Bonn
Bonn, Germany

Katharina Kieslich
Department of Political Science
University of Vienna
Vienna, Austria

Peter Dabrock
Department of Systematic Theology II
University of Erlangen-Nuremberg
Erlangen, Germany

ISSN 2211-4548   ISSN 2211-4556 (electronic)
SpringerBriefs in Philosophy
ISBN 978-3-031-16003-5   ISBN 978-3-031-16004-2   (eBook)
https://doi.org/10.1007/978-3-031-16004-2

This Springer imprint is published by the registered company Springer Nature Switzerland AG
The registered company address is: Gewerbestrasse 11, 6330 Cham, Switzerland

Contents

Chapter 1
Introduction

Abstract The book considers the relationship between governance and participation, and the ways participation has been understood, framed, and applied in the context of synthetic biology (SB) governance. Based on questions about the scope, purpose, and responsibilities assigned to public participation activities, the authors present a literature review of policy reports and articles on SB governance. The brief identifies key characteristics of synthetic biology, such as the complex interplay of scientific, engineering and IT expertise in the field, as well as the ethical and practical challenges these characteristics pose in designing governance frameworks. Drawing on insights from the literature, the authors contest calls for "earlier" and "more" participation because such calls fail to consider the necessary structural adjustments and resources. They offer an approach and recommendations for improving participatory governance in SB by considering questions about skills, training, organization, finances, and normative principles that researchers and policymakers need to consider when designing participation activities. The analytical framework has potential applications in other areas, including governance issues raised during health crises such as pandemics.

Keywords Synthetic Biology · Emerging Biotechnologies · Governance · Participation · COVID-19 · Pandemic

Researchers around the world worked tirelessly to find a vaccine against COVID-19, the disease caused by the coronavirus SARS-CoV-2 that still influences everyday life. The mRNA-based synthetic vaccine[1] produced by Pfizer-BioNTech was the first COVID-19 vaccine and third synthetic RNA therapeutic approved by the FDA (FDA 2021; Pascolo 2021). It was authorized for emergency use a year after the outbreak[2], and, along with other vaccine products, has saved millions of lives (Watson et al.

---

[1] The mRNA-vaccine involves the synthetic replication of mRNA (messenger ribonucleic acid) that gets injected into the body and trains the immune system to build antibodies once SAR-CoV-2 proteins are detected, rather than injecting a dose of antibodies upfront.

[2] The World Health Organization (WHO) declared the COVID-19 outbreak a public health emergency of international concern on January 30, 2020. The vaccine was first available under emergency use authorization in the UK since December 2, 2020 and since December 11, 2020 in the US. The FDA authorized the marketing of the Pfizer-BioNTech COVID-19 vaccine on August 23, 2021.

© The Author(s), under exclusive license to Springer Nature Switzerland AG 2023   1
S. Siewert et al., *Synthetic Biology and the Question of Public Participation*,
SpringerBriefs in Philosophy, https://doi.org/10.1007/978-3-031-16004-2_1

2022)[3]. Hopes are high for synthetic biology (SB) to be one of the most promising technologies in diagnostics and therapeutics to fight the coronavirus and future infectious diseases (EU 2020; Khan et al. 2022; OECD 2021; Pascolo 2021; The Guardian 2021; WEF 2021).

Renewed attention for synthetic biology, and mRNA therapies in particular (The Economist 2021), has once again evoked questions about the potential risks and uncertainties in the field (Li et al. 2021). The coronavirus pandemic fostered developments in gene technology and, more specifically, synthetic biology in unforeseen ways. In the initial phases of the pandemic, studies on the public response to COVID-19 measures and therapeutics were limited (either regarding scope, medium or area). Research on public attitudes towards vaccines often focuses on sentiment and acceptance (Hu et al. 2021; Paul et al. 2022), vaccination policy (mandatory or optional), misinformation and debates about potential side effects rather than the risks of pathogen-related research (Chandrasekaran et al. 2022). The first mentioned research, whilst important, falls short of considering the constituting processes aimed at involving the public in the development of new biotechnologies.

We argue that in times of crises it is important to uphold standards of transparency and to ensure public participation when it comes to the potential benefits and harms of synthetic biology to guarantee a democratically legitimized process and to safeguard trust in research institutions. While the ethical and social issues at stake have taken on a new urgency during the coronavirus crisis—not least with regard to a sustainable and fair global distribution of therapeutics—they are by no means new to the field of synthetic biology. Rather, they are at the heart of an ongoing debate about emerging biotechnologies and the question of how and when to involve the public(s).

Synthetic biology is a field of biotechnology in which biological, chemical, and engineering processes, knowledge and skills are combined to synthetically create genetic materials, living organisms and biological systems (CBD 2016). It entails a number of social and ethical challenges, such as the relationship between the natural and synthetic, or the understanding of life and living (Braun et al. 2019)[4]. The focus on creating or replicating materials found in nature within the confines of scientific laboratories has led to a persistent view that societies will be skeptical about synthetic biology's potential advances in the fields of medicine, agriculture or food and energy

---

However, the speed of COVID-19 vaccine development can also be attributed to a long history of mRNA vaccine research that facilitated the process (Dolgin 2021).

[3] A more recent study suggests that vaccinations "prevented 14.4 million deaths from COVID-19 in 185 countries and territories between Dec 8, 2020, and Dec 8, 2021" (Watson et al. 2022).

[4] Recent findings in stem cell research have further fueled debates about the ethics of synthetic biology. Scientists at the Weizmann Institute of Science generated synthetic embryos from stem cells ex utero (Tarazi et al. 2022), which, for some, produced concerns about the boundaries of science in creating life. While the prospect of synthetic human embryos might still be far away, the rapid developments in research make it all the more important to engage with ethical questions today.

production. This in turn has led to calls for governance approaches to be more inclusive by creating spaces where publics can join in, contribute to, and participate in the research processes surrounding synthetic biology. Marris (2015) even argues that this view has given rise to a "synbiophobia phobia" in which researchers and policymakers exhibit a fear of public fear, and a perception that the publics' fear presents an obstacle to realizing the promises of synthetic biology research.

While calls for opening research spaces for participatory efforts are not unique to synthetic biology, the field offers an interesting case to study the complex interactions between research, governance and public participation in emerging biotechnologies. In this brief, we argue that current governance approaches to synthetic biology fall short of providing practical orientations for researchers and policymakers when it comes to designing participatory activities. Drawing on insights taken from the literatures on the governance of synthetic biology, as well as governance and public participation in public policy, we offer 10 recommendations that provide guidance for those who design, or advocate for, participatory approaches in synthetic biology.

Initial efforts in the field focused on health and medical applications of synthetic biology technologies, for example DNA vaccines and DNA-based gene therapy products. Currently, SB is used by a number of companies to create non-health related products, such as synthetic vanillin or synthetic leather. In general, and adding to the complexity of communication efforts, there is a broad spectrum of positions within the SB community, ranging from academic research to civil society actors, biodefence, the industry and DIY practitioners (Novossiolova et al. 2021). The perspectives of stakeholders on ethical implications of SB are equally diverse. While civil society organizations, such as the ETC-Group, caution against an industry-driven revolution and catastrophic consequences for conservation (2012, 2018), others, early on, identified SB as the next global health innovation (Rooke 2013).

In the COVID-19 pandemic, SB processes played a role in the race to develop a vaccine to prevent the disease, or to minimize the risk of serious illness. These examples illustrate the diversity of areas in which SB knowledge, skills and processes are applied, but once one begins to uncover the different contexts of these areas, a complex picture emerges with regard to designing more participatory governance approaches. Here, it is important to take a step back to consider the relationship between governance and participation, and the ways participation has been understood, framed and applied in SB governance. Basic but fundamental questions emerge: What is the purpose of participation? Who is responsible and who is held accountable for an engaging and effective participation? How can we provide the appropriate resources? And will participation change things for the better? While these questions concern the entire spectrum of the life sciences (and beyond), the interdisciplinary nature and manifold areas of application of SB pose particular challenges.

The number of potential applications in health care, in the biotechnology industry and in research call for an understanding of governance that takes into account context-specific uncertainties. Researchers (e.g. Tait 2009, 2012) and (government) commissions (IRGC 2010; PCSBI 2010) have emphasized governance approaches that can adapt to the opportunities emerging from new scientific discoveries while being aware of the potential risk for humans and the environment, as well as the interests and values of the stakeholders. Along these lines, current approaches in SB governance call for a more dynamic "art" of governance, understood as a "direct political engagement with non-knowing" that does "not rely so much on a chain-of-command or treaty-based establishment, but may be promoted through a transnational joint production of 'orientational questions'" (Zhang et al. 2011, 24).

However, the proclamational nature of those more dynamic governance approaches outruns the structural realities of the research environment. Especially when we perceive of governance not only as an "embodiment of reflexiveness, but the *continuous provocation* of reflexiveness among stakeholders" (Zhang et al. 2011, 24, emphasis added), we need to ask: What are the (structural) prerequisites to enable and facilitate this specific understanding of governance? We argue that participation should be considered at different levels of governance—i.e. hard law, soft law, education, research impact, and research infrastructure—in order to guarantee a visible, critical, and yet constructive public engagement with SB. Participation is not limited to communicating the output of research projects, but is seen as an *end-to-end principle*. We propose six focus areas for these levels of governance that help to identify prerequisites and facilitate elements of genuine and effective public engagement.

In this brief, we build on extensive reviews of the literatures on SB governance, as well as on governance and public participation in public policy, to underpin the argument that a focus on six areas—namely representation, communication, evaluation, resources, training and data management—is necessary to provide more specificity and feasibility to SB governance frameworks that call for 'more', 'earlier' or 'better' participation. In doing so, we hope to provide several practical orientations for researchers, regulators, industry, civil society organizations and policymakers involved in the governance of SB. We begin by identifying trust and justice as core values that underline calls for public participation in emerging biotechnologies. We then turn to a discussion of the literature on governance and public participation generally, and an overview of the literature on SB governance specifically, to demonstrate the paucity of practical recommendations of participatory efforts in SB. Finally, we discuss the development of a vaccine to prevent COVID-19 as an example that has brought the importance and practical implications of public participation in synthetic biology to the forefront of the science-society interface.

References

Braun M, Fernau S, Dabrock P (2019) (Re-)designing nature? An overview and outlook on the ethical and societal challenges in synthetic biology. Adv Biosyst 3(6):1800326

Chandrasekaran R et al (2022) Examining public sentiments and attitudes toward COVID-19 vaccination: infoveillance study using twitter posts. JMIR Infodemiology 2(1):e33909. https://doi.org/10.2196/33909

Convention on Biological Diversity (CBD) (2016) Conference of the parties to the convention on biological diversity, Thirteenth Meeting, Cancun, Mexico. Decision adopted by the conference of the parties to the convention on biological diversity, XIII/17, Synthetic Biology. Retrieved from https://www.cbd.int/doc/decisions/cop-13/cop-13-dec-17-en.pdf

Devlin H (2021) Flu, cancer, HIV: after Covid success, what next for mRNA vaccines? The Guardian. Retrieved from https://www.theguardian.com/science/2021/nov/01/flu-cancer-hiv-after-covid-success-what-next-for-mrna-vaccines

Dolgin E (2021) The tangled history of mRNA vaccines. Nature 597(16):318–324

ETC Group, Friends of the Earth, CTA (2012) The principles of oversight for synthetic biology. ETC Group. Retrieved from https://www.etcgroup.org/sites/www.etcgroup.org/files/The%20Princip les%20for%20the%20Oversight%20of%20Synthetic%20Biology%20FINAL.pdf

ETC Group and Heinrich Böll Stiftung (2018) Forcing the farm: how gene drive organisms could entrench industrial agriculture and threaten food sovereignty. ETC Group. Retrieved from https://www.etcgroup.org/content/forcing-farm

European Commission (2020) EU horizon 2020 on public engagement in responsible research and innovation. Publication office. Retrieved from https://ec.europa.eu/programmes/horizon2020/en/h2020-section/public-engagement-responsible-research-and-innovation

FDA (2021) FDA news release. FDA approves first covid-19 vaccine. Approval signifies key achievement for public health. Retrieved from https://www.fda.gov/news-events/press-announ cements/fda-approves-first-covid-19-vaccine

Hu T et al (2021) Revealing public opinion towards COVID-19 vaccines with twitter data in the united states: spatiotemporal perspective. J Med Internet Res 23(9):e30854. https://doi.org/10.2196/30854

International Risk Governance Council (IRGC) (2010) Guidelines for the appropriate risk governance of synthetic biology. IRGC. Retrieved from https://irgc.org/wp-content/uploads/2018/09/irgc_SB_final_07jan_web.pdf

Khan A, Ostaku J, Aras E, Seker U (2022) Combating infectious diseases with synthetic biology. ACS Synth Biol 11(2):528–537

Li J et al (2021) Advances in synthetic biology and biosafety governance. Front Bioeng Biotechnol 9:598087. https://doi.org/10.3389/fbioe.2021.598087

Marris C (2015) The construction of imaginaries of the public as a threat to synthetic biology. Sci Cult 24(1):83–98. https://doi.org/10.1080/09505431.2014.986320

Novossiolova T et al (2021) Addressing emerging synthetic biology threats: the role of education and outreach in fostering effective bottom-up grassroots governance. In: Trump et al (eds) Emerging threats of synthetic biology and biotechnology. NATO Science for Peace and Security Series C: Environmental Security, pp 86–102. https://doi.org/10.1007/978-94-024-2086-9_6

Organization for Economic Cooperation and Development (OECD) (2021) Accelerating innovation to meet global challenges: the role of engineering biology. OECD Science, technology and innovation outlook 2021: times of crisis and opportunity. OECD Publishing. Retrieved from https://www.oecd-ilibrary.org/science-and-technology/oecd-science-technology-and-innovation-outlook-2021_c1aaa84d-en

Pascolo S (2021) Synthetic messenger RNA-based vaccines: from scorn to hype. Viruses 13:270. https://doi.org/10.3390/v13020270

Paul KT et al (2022) Anticipating hopes, fears and expectations towards COVID-19 vaccines: a qualitative interview study in seven European countries. SSM Qual Res Health 2:100035. https://doi.org/10.1016/j.ssmqr.2021.100035

Presidential Commission for the Study of Bioethical Issues (PCSBI) (2010) New directions: the ethics of synthetic biology and emerging technologies. Retrieved from https://bioethicsarchive.georgetown.edu/pcsbi/sites/default/files/PCSBI-Synthetic-Biology-Report-12.16.10_0.pdf

Rooke J (2013) Synthetic biology as a source of global health innovation. Syst Synth Biol 7:67–72. https://doi.org/10.1007/s11693-013-9117-3

Tait J (2009) Governing synthetic biology: processes and outcomes. In: Schmidt M et al (eds) Synthetic biology: the technoscience and its societal consequences. Springer, pp 141–154

Tait J (2012) Adaptive governance of synthetic biology. EMBO Rep 13(7):579–579. https://doi.org/10.1038/embor.2012.76

Tarazi S et al (2022) Post-gastrulation synthetic embryos generated ex utero from mouse naive ESCs. Cell 185(18):3290–3306

The Economist (2021) Covid-19 vaccines have alerted the world to the power of RNA therapies. The Economist. Retrieved from https://www.economist.com/briefing/2021/03/27/covid-19-vaccines-have-alerted-the-world-to-the-power-of-rna-therapies

Watson OJ et al (2022) Global impact of the first year of COVID-19 vaccination: a mathematical modelling study. Lancet 22(9):1293–1302. https://doi.org/10.1016/S1473-3099(22)00320-6

World Economic Forum Global Future Council on Synthetic Biology (WEF) (2021) Revisiting and realizing the promises of synthetic biology. Briefing paper. WEF. Retrieved from https://www3.weforum.org/docs/WEF_Revisiting_and_Realizing_the_Promises_of_Synthetic_Biology_2021.pdf

Zhang JY, Marris C, Rose N (2011) The transnational governance of synthetic biology: scientific uncertainty, cross-borderness and the 'art' of governance. BIOS working paper no. 4. BIOS, London School of Economics and Political Science. Retrieved from http://royalsociety.org/uploadedFiles/Royal_Society_Content/policy/publications/2011/4294977685.pdf

Chapter 2
Imagined Futures: Trust and Justice in Synthetic Biology Governance

Abstract The chapter discusses social and cultural implications and current areas of conflict in synthetic biology. Since there is the challenge of potentially misleading imaginaries and an overall uncertainty regarding research and application in the field, it is quintessential to consider the category of social trust as an underlying value that shapes our perception and acceptance of SB. Trust in synthetic biology cannot simply be guaranteed or established, but remains fragile. At the same time, there are conditions which make a leap of faith more likely. A central component of this is participation and a structure of governance that is aligned to an understanding of the complexity of negotiations between science, society, and economy.

Keywords Risks · Precaution · Imaginaries of Synthetic Biology · Bioeconomy · Trust

There are areas of conflict in synthetic biology that future efforts in participation will have to address. First and foremost, there seems to be a significant gap between the anticipated potential and risks associated with synthetic biology, its estimated market value (and distribution among different stakeholders, e.g. federal and private product research) and its status quo. While the global synthetic biology market for products is expected to grow from \$9.5 billion in 2021 to \$33.2 billion by 2026 with a compound annual growth rate (CAGR) of 28.4% (BBC Research 2022), synthetic biology has still some way to go considering the technical constraints that limit data collection and analyses and the unreliability of output. Machine learning approaches are slowly generating more robust models of biological systems that will, in the long run, improve experimental design for future engineering strategies. Most experiments, however, are still based on a trial-and-error procedure (Karoui et al. 2019). The research impact and future market of SB also raises questions concerning the fair and equitable sharing of benefits within the global bioeconomy. All stakeholders in SB should consider the distributional justice questions from the perspective of their particular space of enunciation. Issues of inequality demand a structured dialogue that precedes and extends specific participatory modes and call for a broader awareness on all levels of governance.

In general, social and cultural implications have moved to the center of attention, emphasizing the role of different stakeholders, communities and publics for the design of SB governance (Cummings and Kuzma 2017; ETC 2013; European Commission 2010; IUCN 2019; NASEM 2016; OECD 2014; Stirling et al. 2018; UNEP 2019; Wallach et al. 2018; Zhang et al. 2011). This shift ties in with broader discussions on the societal resonance of biotechnologies, prominently addressed by Jasanoff and Kim (2009, 2015), who have emphasized the importance of socio-technical imaginaries as collective visions of good and attainable futures. While this provides an important argument to unlock the potential of public participation, other critics have cautioned against the consequences of such imaginaries and their techno-utopian elements, which potentially distort the image of synthetic biology in the public and regulatory eye. In fact, "myths", fueled by techno-fictions, the emergence of do-it-yourself communities and the potential implications of dual-use-research have shaped the "techno-epistemic culture of synthetic biology" where "the tension between the possible and the actual is a criterion of distinction between the various trends that constitute synthetic biology" (Bensaude-Vincent 2013). Jefferson et al. state that public discourse on synthetic biology and biosecurity "tends to portray speculative scenarios about the future as realities in the present or the near future" (2014). Furthermore, these myths define understandings of synthetic biology and shape policy narratives. The authors argue that current debates often focus on access to biological materials and digital information, yet it would be more productive to look at tacit knowledge that evolves from the interaction of different types of expertise and to shift attention to human practices and institutional dimensions.

Another issue pertains to the uncertain nature of the field. Some of the impacts of synthetic biology and its biological and socio-economic benefits and risks are not yet foreseeable or predictable. Risk assessment has to be adjusted repeatedly due to the changing nature of the field itself—and even more importantly—due to changing socio-economic conditions and perceptions of synthetic biology. Hence, a purely technical risk assessment proves to be inadequate. The very emphasis on risk and the precautionary principle in the context of synthetic biology is under scrutiny and challenged by cost–benefit approaches, often presented within a broader scope, e.g. with an emphasis on values or/and the science-policy interface (Wallach et al. 2018).

In this light, the notion of precaution has been reframed as a "sensitive and innovation-driven approach" that might be able to resolve the vagueness of emerging biotechnologies (Braun and Dabrock 2018). Such a vagueness can be described as an uncertainty within existing regulation frames regarding the future development of an emerging technology such as synthetic biology. A regulation framework as mentioned before is fundamentally in need of a robust setting of scientific as well as governmental institutions that are able to supply a framework capable of facing the speed of technological advancement and still provide orientation. Although providing such an orientation is a difficult task per se, the current institutional setting is further-more challenged by another development: While institutions are locally and nation-ally embedded, they are also confronted with a global scientific development where a scientific endeavour in other regions with a different research agenda has a massive impact on the local institutional setting.

To deal with the described vagueness in a manner that is both theoretically ambitious and practically viable, a multidimensional methodology is required. This includes a so-called first wave ethical assessment: the evaluation and weighing up of the intentions, prospects and risks of a certain biotechnology (in this case, of synthetic biology). Within such a first wave framing of potential risk, there is an increasing need for corresponding measures to be taken, as proposed by the German Ethics Council (among others) in its biosecurity opinion issued in 2014.

The important and hitherto unsolved problem in dealing with the vagueness of emerging biotechnologies is how we can define and subsequently implement over-arching rules of conduct for a common international safety culture which addresses both the researchers which are institutionally embedded as well as those who are not (Trevan 2015). Thus, the aim must be to reinforce the position of labs and research institutions as highly reliable and trustworthy organizations in order to hold on to the trust placed in these institutions by the broader public (Siewert and Dabrock 2020). It is important to keep in mind that, while this trust exists, it remains fragile (Braun et al. 2015a, b). Even though it is necessary to engage in a broader debate about the possible basic principles for such a safety culture, of the key elements involved, the first should be a focus on the aims which science should pursue within different fields of application. The second priority should be the development of appropriate codes of conduct, and the third the establishment of international commissions which both monitor the processes used and help to improve the relevant codes of conduct. For the robustness of a safety culture as described before, it is crucial that the imple-mentation of such a strategy is to be understood as an institutional arrangement and becomes an integral part of routine daily actions.

As a proactive measure, it is important to promote further self-reflexive critique in ethical exploration. Such a self-reflexivity can be successful if (as much and as intensively as possible) an upstream process of dialogue with the public is sought through a process that is both transparent and participatory (demonstrating public engagement *with* science). Such an endeavor requires a certain competence on the part of those initiating public engagement activities. But it is precisely in these communications and public negotiation processes that it becomes challenging when the subject of the research is either quite abstract or requires a high degree of detailed knowledge.

However, in order to avoid misconceptions, normative questions have to be tied back to the actual state of research and more practical concerns of those working in the lab and those who will be affected outside of it—even though it requires bigger efforts in conceptualizing participation and in translating scientific knowledge. To put it more bluntly: A framework of trust and justice requires a reality check with everyone involved. A target-oriented communication of the specifics and actual results of research is the foundation of an effective activation of the public.

In this way, this approach aims to confront the critical questions of biotechnology at the outset with publicly-expressed hopes, expectations and anxieties actively helping to shape the process of research decision-making on the basis of a broad range of public opinions. At the same time, it would be a fallacy to assume that it is possible to solve ethical problems primarily with procedural strategies. Formal stan-dards are necessary, but by no means sufficient as conditions of ethical reflection. Especially in recent times, the ideal picture of broadening public participation as much as possible has begun to crack (Prainsack 2014, 2017). However, the converse point is also valid: Ethical reflections remain without effect unless corresponding criteria-led considerations are developed alongside strategies that are sensitive to how they are likely to be received.

A responsible and trustworthy governance that keeps emphasizing the connections between grand challenges and specific issues might enable a spirit of creative anticipation and willingness to partake in imagining the future. This, in turn, can become a guiding principle of policy and instrument of its legitimation. Participation offers a way to enable a self-determined response to emerging technologies and thus provides an important tool to foster trust in SB research and applications. However, it is important to keep in mind the fragile nature of trust that is always based on "a relational mode of interaction that can in principle establish or enhance room for maneuver", so that even distrust "has a fruitful function in that it helps to indicate emerging shifts in power, justification, or representation" (Braun et al. 2021).

In order to advance a strategy of effective Responsible Research and Innovation (RRI) governance which functions as more than window dressing, a third branch of responsible ethical reflection has been developed at the threshold of science and society. Empirical studies demonstrate that many people who welcome scientific progress have a very limited understanding of the enormous level of complexity of many research endeavours (European Commission 2014). Therefore, building trust in research projects often has to be shifted into a form of trust-building activity for institutions. Whereas, generally speaking, public institutions engender more trust than private organizations, the lines between public or basic research and development on the one side and private research and development on the other side are increasingly blurred. Moreover, the political framing in the EU and other countries also encourages publicly funded institutions to generate spin-offs and collaborate with companies in the respective field.

Taking into account that public trust is often effectively granted more like a blank cheque, it remains precarious, since it is quickly withdrawn in cases of misuse or scandal. Whole branches of research, and not only the wrongdoer, can be affected by these incidents. In view of current challenges to democratic and civil society consensus-building, it seems enormously important whose knowledge and what level of knowledge is recognized as expertise. At the same time, however, the recent history of biotechnological innovations and developments shows that disappointed trust in the face of the actions of individuals or institutions can lead to mistrust of an entire body of research. Such distrust would be misconceived if it were understood as merely the opposite of trust. On the one hand, mistrust can mean that trust has turned into distrust and that trust has been withdrawn from an entity, an institution or even a development. On the other hand, distrust also fulfills an important critical function in civil society: To whom are the gains and to whom are the risks of a technological development attributed? Is the attributed expertise really available, or is there a need for further forms of expertise as a basis for public consensus?

In this sense, putting trust in a technology, an institution, or certain processes could be understood as a wager between science, economy and society. The goal of such a process is to arrive at conditions in which for those affected, trust and/or reliance is a good bet. Nevertheless, understanding trust and/or reliance as flowing from this kind of negotiation process is only one way in the understanding of trust. By giving trust or deciding to rely on an entity or institution, a specific relation between the person and the respective entity or institution is initiated. As some of us have argued, putting trust in an entity involves an element of risk, the willingness to get involved and commit to something, or in other words: giving a leap of faith (Braun et al. 2021).

Of course, certain background conditions matter for whether or not such leaps occur. For societal trust in a new technology, for example, it matters that development processes are open, inclusive, and transparent. But given the foregoing, such conditions might not be sufficient. And even then, the endowment granted through a leap of faith remains fragile (Braun and Meacham 2019). For example, once violations of the trustor's goodwill are perceived, trust may be withdrawn. This is a live possibility for synthetic biology, where ambiguous antecedent perceptions could easily tip towards skepticism, disappointment and resistance unless actual reservations, worries, and potentially destabilizing factors surrounding the technology are taken seriously (Trump et al. 2021).

A central building block for responsible research and innovation in the field of synthetic biology is social trust in the persons and institutions involved. This cannot simply be guaranteed or established, but remains fragile. At the same time, however, there are conditions which, if fulfilled, make it more likely that a leap of faith will be made. A central component of this is social participation. Such participation comprises two levels: The first is the central question of how profits and risks are distributed. The second concerns the question of what forms of concrete participation exist.

Participation can be a productive avenue to foster trust, even more so when participatory strategies are embedded in reliable governance structures. Before we can explore the prerequisites and the infrastructure of effective participatory activities in synthetic biology, we need to understand how governance has been conceptualized to deal with complex technological, ethical and social questions. Hence, in the next section, we discuss the challenges in designing governance frameworks generally, and for synthetic biology particularly.

References

BBC Research (2022) Synthetic biology: global markets. Retrieved from https://www.bccresearch.com/market-research/biotechnology/synthetic-biology-global-markets.html

Bensaude-Vincent B (2013) Between the possible and the actual: philosophical perspectives on the design of synthetic organisms. Futures, Elsevier 48:23–31

Braun M, Starkbaum J, Dabrock P (2015a) Safe and sound? Scientists' understandings of public engagement in emerging biotechnologies. PLoS ONE 10(12):e0145033. https://doi.org/10.1371/journal.pone.0145033

Braun M, Starkbaum J, Dabrock P (2015b) The synthetic biology puzzle: a qualitative study on public reflections towards a governance framework. Syst Synth Biol 9(4):147–157

Braun M, Dabrock P (2018) Mind the gaps! Towards an ethical framework for genome editing. EMBO Rep 19(2):197–200. https://doi.org/10.15252/embr.201745542

Braun M, Meacham D (2019) The trust game: CRISPR for human germline editing unsettles scientists and society. EMBO Rep 20:e47583. https://doi.org/10.15252/embr.201847583

Braun M, Bleher H, Hummel P (2021) A leap of faith: is there a formula for "trustworthy" AI? Hastings Cent Rep 51(3):17–22. https://doi.org/10.1002/hast.1207

Cummings CL, Kuzma J (2017) Societal risk evaluation scheme (SRES): scenario-based multi-criteria evaluation of synthetic biology applications. PLoS ONE 12(1):e0168564. https://doi.org/10.1371/journal.pone.0168564

El Karoui M, Hoyos-Flight M, Fletcher L (2019) Future trends in synthetic biology—a report. Front Bioeng Biotechnol 7:175. https://doi.org/10.3389/fbioe.2019.00175

ETC Group (2013) Synthetic biology: the bioeconomy of landlessness and hunger. ETC Group. Retrieved from http://www.etcgroup.org/issues/synthetic-biology.

European Commission, European Group on Ethics in Science and New Technologies (2010) Ethics of synthetic biology. Publication Office. Retrieved from https://data.europa.eu/doi/https://doi.org/10.2796/10789

European Commission (2014) Special Eurobarometer 419: public perceptions of science, research and innovation. Publication Office. Retrieved from http://ec.europa.eu/public_opinion/archives/ebs/ebs_419_en.pdf

International Union for Conservation of Nature (IUCN) (2019) Genetic frontiers for conservation: an assessment of synthetic biology and biodiversity conservation. Synthesis and key messages. IUCN. Retrieved from https://www.iucn.org/resources/publication/genetic-frontiers-conservationsynthesis-and-key-messages

Jasanoff S, Kim S-H (2009) Containing the atom: sociotechnical imaginaries and nuclear power in the United States and South Korea. Minerva 47(2):119–146

Jasanoff S, Kim S-H (eds) (2015) Dreamscapes of modernity: sociotechnical imaginaries and the fabrication of power. University of Chicago Press

Jefferson C, Lentzos F, Marris C (2014) Synthetic biology and biosecurity: challenging the "myths." Front Public Health 2:115. https://doi.org/10.3389/fpubh.2014.00115

Organization for Economic Cooperation and Development (OECD) (2014) Emerging policy issues in synthetic biology. OECD Publishing. Retrieved from https://doi.org/10.1787/9789264208421-en

Prainsack B (2014) Understanding participation: the 'citizen science' of genetics. In: Prainsack B, Schicketanz S (eds) Genetics as social practice. Ashgate, pp 147–164

Prainsack B (2017) Solidarity in biomedicine and beyond. Cambridge University Press

Siewert S, Dabrock P (2020) Pluralität achten – Nachdenklichkeit erzeugen – Orientierung anbieten: Der Deutsche Ethikrat als Spiegel und Katalysator gesellschaftlicher Debatten. Health Academy 3. In: Manzeschke A, Niederlag W (eds) Ethische Perspektiven auf Biomedizinische Technologie. De Gruyter, pp 246–261. https://doi.org/10.1515/9783110645767-023

Stirling A, Hayes KR, Delborne J (2018) Towards inclusive social appraisal: risk, participation and democracy in governance of synthetic biology. BMC Proc 12(Suppl 8):15. https://doi.org/10.1186/s12919-018-0111-3

The National Academies of Sciences, Engineering, and Medicine (NASEM) (2016) Gene drives on the horizon: advancing science, navigating uncertainty, and aligning research with public values. The National Academies Press. Retrieved from https://nap.nationalacademies.org/catalog/23405/gene-drives-on-the-horizon-advancing-science-navigating-uncertainty-and

Trevan T (2015) Rethink biosafety. Nature 527:155–158

Trump BD et al (eds) (2021) Emerging threats of synthetic biology and biotechnology: addressing security and resilience issues. NATO Science for Peace and Security Series C: Environmental Security. https://doi.org/10.1007/978-94-024-2086-9_1

United Nations Environement Programme (UNEP) (2019) Frontiers 2018/2019: emerging issues of environmental concern. United Nations Environment Programme. Retrieved from https://www.unep.org/resources/frontiers-201819-emerging-issues-environmental-concern

Wallach W, Saner M, Marchant G (2018) Beyond cost-benefit analysis in the governance of synthetic biology. Governance of emerging technologies: aligning policy analysis with the public's values. Hastings Cent Rep 48(1):S70–S77. https://doi.org/10.1002/hast.822

Zhang JY, Marris C, Rose N (2011) The transnational governance of synthetic biology: scientific uncertainty, cross-borderness and the 'art' of governance. BIOS working paper no. 4. BIOS, London School of Economics and Political Science. Retrieved from http://royalsociety.org/uploadedFiles/Royal_Society_Content/policy/publications/2011/4294977685.pdf

Chapter 3
Approaches in Defining Governance

Abstract The chapter discusses how governance in emerging biotechnologies has been conceptualized to deal with complex technological, ethical, and social questions. The authors explore the challenges in designing governance frameworks generally and for synthetic biology particularly. An understanding of different approaches in governance, their advantages and disadvantages, can help us draw a more precise picture of what is being governed, by whom and how in the field of synthetic biology.

Keywords Governance · Complexity · Deliberation · Decision-Making · Public and Private Actors

3.1 Designing Governance Frameworks

Governance is a term that is often used but rarely defined. It is used as a concept to describe how something—a policy field, a branch of research, an industrial sector—is managed, organized or regulated. Alternatively, it is used to describe the changing patterns of government activities, where formerly public spheres of activity are outsourced to private or quasi-public entities. Given these varied uses of the concept of governance, it is surprising that a discussion of the scope of the term is rarely included in governance reports and frameworks. The reason for this may be as simple as assuming that stakeholders in a given field know what is meant when applying the term, or the reason may be rooted in the pragmatic consideration of the need to avoid further complexity in complex subject areas. The validity of these possible explanations notwithstanding, a short excursion on the varied applications of governance is a helpful basis from which to develop recommendations for the governance of any sector, let alone a complex and innovative one such as synthetic biology. The excursion helps us answer important governance questions, such as *what* is being governed, by *whom and how*. These questions may appear simple, but as the next paragraphs show, both the academic literature as well as extant governance frameworks rarely explore even these most basic questions.

A challenge in defining governance, and in outlining demarcations around its usage, is that there is not one particular literature to which one can turn. Governance branches many fields, different literatures, and diverse areas of public and private life. The following sections should therefore not be read as an exhaustive discussion of governance theories, concepts and applications, but as a starting point from which to build governance frameworks for emerging biotechnologies. In political science and public policy, for example, governance is used as an umbrella term to describe the changes that many governments have undergone, or have undertaken, in the past decades. These changes encompass modifications to government styles, such as the opening up of spheres of activities traditionally thought of as being in the purview of governments to other entities such as quasi-public or private entities. Common explanations for these changes include the popularity of efficiency-driven approaches to the management of public activities such as new public management (NPM), which promotes the idea that government activities can be run more efficiently, and cost effectively, if they are run more like private organizations. Another explanation for the change in government styles, however, is the idea that ever more complex policy questions demand a more comprehensive involvement by society and stakeholders. In this view, the responsibility for decision-making becomes more dispersed, and spaces open up for the public to become involved in decision-making. This point is particularly relevant for the governance frameworks of science and research, and this report takes it up again in a later section.

Along the lines above, Rhodes defines governance as "[…] a change in the meaning of government, referring to a new process of governing; or a changed condition of ordered rule; or the new method by which society is governed" (1996, 652–653). Stoker builds on the definition by saying that governance connotes a difference in processes in how government matters are handled or decided upon. He goes on to say: "There is, however, a baseline agreement […] that governance refers to the development of governing styles in which boundaries between and within public and private sectors have become blurred" (1998, 15). By extension, the responsibilities for societal and economic action become blurred and dispersed between different actors, not just government actors.

In addition to the changes in meanings of governments and processes of decision-making, others point out that governance is characterized by "ways of governing that are non-hierarchical, involve networks of actors, both public and private, determining policy (and regulation) through negotiation, bargaining and participation" (Weale 2011, 58). It follows that governance is marked by processes of conversation, discussion and deliberation that are different from traditional top-down formats of governmental decision-making. In designing governance frameworks it is therefore essential to consider possible spaces (physical and otherwise) that will open up processes of decision-making to accommodate more interactive styles of communication.

The list of definitional elements of governance is long. Greer et al. (2016) point out that governance concepts usually contain normative lists of values and attributes— good governance, scientific governance, research governance, public governance, deliberative governance, to name a few—that are arbitrary in their selection. In our view, the arbitrary assignment of values and attributes to different forms of governance leads to a lack of clarity on the purpose that the governance concept serves in individual settings. Using governance with an adjective ('good', 'deliberative', 'discursive') is different from using governance to describe the way in which a field, a sector, a research area, is structured and managed. The first carries normative connotations as to the quality of governance, whereas the second implies the need to make practical decisions with regard to designing governance structures. Both elements have intrinsic value and naming them allows stakeholders to distinguish between normative and practical aspirations of governance.

3.2 Governance of Synthetic Biology

There are a number of elements from the above excursion that are particularly relevant when designing a framework for emerging biotechnologies such as synthetic biology. First, governance approaches acknowledge that there is something inherently complex about the sector, the field, the question that is to be governed. Second, the governance literature acknowledges that decision-making and agenda-setting stand to gain from bottom-up rather than top-down approaches. Third, governance approaches place particular emphasis on the networks of public and private actors that make decisions through deliberative and discursive spaces. Rather than taking these three elements for granted, however, a holistic approach to the governance of emerging technologies demands that one examines the ways in which each of these three elements are relevant in a given area for which governance frameworks are proposed.

Position papers and reports on synthetic biology (see for example DFG 2018; EASAC 2010) commonly provide explanations for why the governance of synthetic biology is particularly complex, complexity being the first of the elements identified in the discussion of the literature on governance. The explanations include the fact that synthetic biology, as a research field, is multidisciplinary rather than being confined to one branch of science; that the 'outcome', the products, of synthetic biology research are not easily named because much uncertainty still exist around future applications, which has implications for intellectual property rights (IPR); and that many of the benefits, as well as the risks, of synthetic biology are yet unknown, which in turn has implications for biosafety and biosecurity considerations. In other words, synthetic biology exhibits characteristics that give rise to complexity, which in turn demands governance. Lowi (1964) referred to such characteristics as *issue characteristics* and posited that the nature of certain policy areas or issues of public life are more complex than others, for example by giving rise to challenging ethical and moral questions. Different governance styles follow the different characteristics of issues, according to Lowi. The characteristics provide an answer to our original questions of *what* needs to be governed.

Often touched upon, but rarely elaborated, we add issues of public or societal value, innovation, and timeline considerations to the list of issues that makes synthetic biology a complex emerging biotechnology. These concepts are notoriously hard to define, value-laden and controversial terms. Calls for the outcomes of science endeavours to be of public value, and for innovation not to be stifled through overly burdensome regulation, are common when talking about the governance of science.

However, public value and innovation are not neutral terms, and much like governance they are rarely defined, meaning that stakeholders will work towards their own visions and versions of these concepts. This is exacerbated by the fact that the timeline for innovations in synthetic biology is uncertain, meaning stakeholders are faced with governing processes and outcomes that they cannot name, let alone ascribe a timeline to them. That is not to say that we should strive towards all-encompassing definitions of difficult concepts to which we can all subscribe, but that we need to be aware that the understandings of these terms will need to be negotiated, and re-negotiated, as emerging biotechnologies develop. Deliberative spaces that allow for iterative discussions and negotiation therefore acquire a special standing in the governance of such biotechnologies.

The second element of governance is the recognition that decision-making and agenda-setting occurs in formats other than traditional top-down formats. Especially in synthetic biology, where scientific methods and relationships are still emerging, it is important to provide flexible decision-making and agenda-setting structures that allow researchers to take their research into unchartered territory, whilst heeding potential societal and political concerns over what will be found in this territory. To a certain extent, the recognition for more iterative decision-making is reflected in calls for an anticipatory and innovation governance approach to synthetic biology where governance frameworks can be developed and adjusted when innovations and scientific findings necessitate it. In other words, anticipatory governance approaches are an approximation to the question of *how* synthetic biology might be governed.

Third, governance approaches place emphasis on the networks of public and private actors that make decisions through deliberative and discursive spaces. Here, deliberative governance has emerged as a concept to signify the trend of governments, private companies, and scientific communities to engage in more meaningful ways with their constituents (Hendriks 2009). Deliberative governance offers insights into the question of *who* governs in a particular sector. It is rooted in theories of deliberative democracy that promote the idea that people who are affected by decisions (policy decisions or otherwise) should be involved in making those decisions.

Several aspects directly affect the choice of adequate public engagement strategies: What part(s) of synthetic biology might be governed? How can they be governed and by whom and when? It is crucial to look at the ways participation has been understood in governance approaches on synthetic biology in order to identify future pathways. We conducted an extensive literature review that aims to provide answers to the following questions: What is the particular value ascribed to participation in major policy reports and articles on SB? How is participation defined, framed and perceived in relation to the idea of governance? What role does participation play at different governance dimensions, from hard law to education? What is the level of analysis and consideration with respect to the variety of public engagement designs and purposes? The next section does not provide a conclusive survey of existing perspectives, but offers a summary of our findings, while highlighting certain developments and trends of a potential 'governance though participation' approach.

References

Deutsche Forschungsgemeinschaft (DFG) (2018) Synthetische Biologie. Standortbestimmung. DFG. Retrieved from https://www.dfg.de/download/pdf/dfg_im_profil/geschaeftsstelle/publikati onen/stellungnahmen_papiere/2018/181008_synthetische_biologie_standortbestimmung.pdf

European Academies Science Advisory Council (EASAC) (2010) Realising European potential in synthetic biology: scientific opportunities and good governance. EASAC. Retrieved from https://www.leopoldina.org/uploads/tx_leopublication/2010_EASAC_Statement_Synthe tic_Biology_ENGL.pdf

Greer S-L, Wismar M, Figueras J (eds) (2016) Strengthening health system governance: better policies, stronger performance. Open University Press

Hendriks C-M (2009) Deliberative governance in the context of power. Policy Soc 28(3):173–184

Lowi T (1964) American business, public policy, case-studies, and political theory. World Politics 16(4):677–715

Rhodes RAW (1996) The new governance: governing without government. Polit Stud 44(4):652–667. https://doi.org/10.1111/j.1467-9248.1996.tb01747.x

Stoker G (1998) Governance as theory: five propositions. Int Soc Sci J 155:17–28

Weale A (2011) New modes of governance, political accountability and public reason. Gov Oppos 46(1):58–80

Chapter 4
Governance and Participation in Policy Literature on Synthetic Biology

Abstract The chapter deals with the different positions in engaging the public(s) in the field of synthetic biology and provides an overview of the most prominent avenues in research, policy, and non-governmental action. Against the notion that participation is merely "management of perceptions" or "informing the public", the authors argue for a holistic perspective in participatory approaches that perceives of informative and deliberative participation as equally valid degrees of activation. Furthermore, the authors question an overall emphasis on either institutional or individual responsibilities in the dialogue with the public.

Keywords Literature Review · Participatory Approaches · Stakeholders · Public Engagement

There seems to be little disagreement over the argument that participation is an important asset and legitimizing aspect in the governance of emerging biotechnologies. In fact, the Economic and Social Research Council has defined participation as one of the essential features of innovation governance along with precaution and responsibility to "'open up' scrutiny and accountability beyond anticipated consequences alone, to also interrogate the driving purposes of innovation. These qualities celebrate that innovation is not a matter for fear-driven technical imperatives, but requires a democratic politics of contending hopes" (Stirling 2015, 1).

Along these lines and specifically tailored to synthetic biology, Wallach, Saner, and Marchant state:

> Any effort to improve the governance of synthetic biology should start with a rich understanding of the different possible science-policy interfaces that could help to inform governance. [...] Forging appropriate means for evaluating societal impacts, to the best of a corporation's, industry's, or government's ability, is central to responsible research and innovation. Complex problems require a variety of expertise and sophisticated forms of teamwork and consultation. Policy-makers and process designers need to focus on the whole system: people, processes, and performance. (2018, 71)

To secure the inclusion of the public in not only commenting but rather in being actively engaged in shaping the discourse of emerging technologies, Trump has argued that the involvement of non-governmental stakeholders provides an important feedback. It might help to "communicate to policy makers and regulatory decision makers with respect to (i) realistic risk and benefit outcomes posed by emerging sciences, (ii) societal perception and response to such sciences, and (iii) aligning incentives and research goals for developers moving forward with respect to various areas of hazard, exposure, and consequence measurement for health risk" (2017, 1143).

While most (and even diverging) governance proposals assume the positive effects of (public) participation (European Commission 2010; Friends of the Earth 2012; IRGC 2010; PCSBI 2010), long-term empirical work on the actual outcome of an embedded participatory approach and the resonance of different instruments in the diverse and specific contexts of synthetic biology is scarce. According to the National Academies of Sciences, Engineering and Medicine (NASEM) there is still a need for a structured and continuous engagement and the establishment of clearer engagement pathways:

> Governing authorities, including research institutions, funders, and regulators, should develop and maintain clear policies and mechanisms for how public engagement will factor into research, ecological risk assessments, and public policy decisions on gene drives. Defined mechanisms and avenues for such engagement should be built into the risk assessment and decision-making processes from the beginning. (NASEM 2016, 10, recommend. 9-5)[1]

Taking a closer look at the ways participation is conceptualized, it is often framed and understood as 'management of perceptions' or as 'informing the public' (see e.g. Anderson et al. 2012; EASAC 2010). But participation is much more than public attitude research aiming to secure public acceptance. As Zhang, Marris and Rose argue: "[D]espite the growing attentiveness to public engagement, practice seems to resemble mostly a variation of broad-base conversation, or worse, public education or public attitude research. There seems to be little evidence that these endeavors, as currently formulated, can contribute to the governance of research practice" (2011, 25). In line with the argument of the authors, the literature review shows that there are only few approaches that actually work with a nuanced and innovative understanding of participation. In the following we will address some of the more prominent and substantive examples.

While not directly aiming at the governance of synthetic biology, the Nuffield Council for Bioethics provides important incentives for emerging biotechnologies by promoting a "public ethics discourse" (2012) that puts deliberation and public exchange at the heart of governance strategies. In this vein, the International Risk Governance Council (IRGC) has argued for a more structured dialogue between risk assessors, scientists, regulators and the full range of stakeholders to identify relevant developments, evaluate the appropriateness of current risk assessment and regulatory frameworks. This would enable a discussion about potential risk management options and their interactions with technology development as a basis on which to build risk governance strategies. The IRGC even proposes to seek new forms of communication and dissemination that reach beyond the traditionally didactic approaches of the

---

[1] While the NASEM report focuses on gene drives, the argument is also applicable to the field of synthetic biology.

'public understanding of science', exploring the interactions of science, technology, policy and society by using theatre, enactments and structured debate. Dialogue around shared interests could lead to more creative regulatory solutions that avoid polarization or "the imposition of a single set of values that unnecessarily constrains the opportunities available to society at large" (2010, 39). Adding to this call, Douglas and Stemerding (2014) also argue for more experiments in anticipatory governance and an open deliberative discourse.

SYNENERGENE, a four-year project supported by the European Commission and established to create an open dialogue on the risks and benefits of synthetic biology, developed context-specific modules and evaluated methods via stakeholder workshops and other forms of participation. The project emphasized governance as a "learning process" and called for open spaces to deliberate on cultural and religious values (rather than opinions). The project partners criticized currents trends, which try to tackle grand challenges, e.g. risk assessment and potential harm from technological and more recently data and AI-driven systems in the field, pushing aside the much needed societal exchange on innovative practice, on entrenched conflicts and wicked problems. In the course of the project, trading zones were organized in Europe and the USA for honest discussion on whether research agendas reflect society's values. A move from risk to innovation governance as envisioned by SYNENER-GENE requires institutional change, in addition to new tools for regulators, as well as the development of open public–private partnerships into public–private-societal partnerships (SYNENERGENE 2017).

More recently, Stirling, Hayes and Delborne emphasized the opportunities and barriers that exist to substantive public engagement and participation in risk-based governance frameworks. The authors analyze different steps wihthin a risk assessment process where public participation would be more or less essential and easy (or hard) to facilitate in view of social appraisal. While they see much potential— also with respect to the diversity of practical participatory methods—they also draw attention to the obstacles of "practical extensions of existing procedures in order to enable meaningful participatory deliberation" (2018, 49). All in all, the authors argue that "questions around benefit and harm must be directed to the potential pros and cons associated with a diverse array of alternative policy options" (47).

Other notable avenues include Cummings' and Kuzma's (2017) Societal Risk Evaluation Scheme (SRES), the application of the TAPIC framework (transparency, accountability, participation, integrity, capacity) (Trump 2017), or Wagner's and Bollaert's extension of the "7P" approach (2013), which defines the seven key intervention points (Principal Investigator, Project, Premises, Provider, Purchaser, Public, and Publisher) among stakeholders to address the governance of synthetic biology (for origins of approach see Kelle 2009 and 2013).

Another very important impulse for the shift towards participatory approaches ties back to a deeper understanding and emphasis of cultural and social implications of SB and a focus on motivations of actors and stakeholders shaping the field. This implies a departure from concerns about 'hard' impacts relating to tangible technological risks for human health and the environment towards 'soft' impacts relating to the complex and often unintended ways in which technologies may transform individual behaviors, public attitudes and values, social relationships, and economic balances of power (Douglas and Stemerding 2014, see also IUCN 2019).

Civil society organizations further strengthen the agency of public actors and different communities that might be affected by synthetic interventions. For instance, the ETC Group, a civil society organization concerned with the impact of emerging technologies on society's most vulnerable members, claims that oversight has to be developed with *full participation* of the public, which means that the public "must have the legally enforceable right to halt dangerous applications, not just comment after decisions have been made". Furthermore, the initiative demands effective participation "throughout the entire decision-making process related to the development of synthetic biology and the products of synthetic biology, including setting the research agenda, the context and the scope of the risk assessment" (2012, 9).

Notwithstanding the importance of a "right-to-know", it remains equally important to be aware of the bias in engagement (IUCN 2019). While there have been concerns voiced by NGOs and scholars about the respective interests of those involved, particularly where engagement is initiated or facilitated by proponents of technology (IUCN 2019), one has to keep in mind that transparency and honest assessment of interests and values must refer to all actors in the field—including non-governmental organizations (Tait 2009).

While Zhang et al. (2011) and others have a point in criticizing a purely deficit-oriented understanding of participation and although the field sees innovative, prospective and inclusive approaches, there seem to be at least three major shortcomings that we want to address. First, even the more dynamic and elaborate conceptions of participation often make use of rather abstract and ideal notions of the concept of "the public" (singular). Second, most of the reports and policy papers focus first and foremost on the governance of science policy regulations (research practice, patents and applications) and thus limit the scope of impact to the realm of soft law (see e.g. Köchy and Hümpel 2012). Third, the responsibility of appropriate and transparent communication is assigned primarily to the individual researchers or (interdisciplinary) research councils (see a.o. DFG et al. 2009; EASAC 2010; Köchy and Hümpel/BBAW 2012; Royal Academy of Engineering 2009a, b, 2011).

With regard to the first objection, we need to ask who we consider to be the public(s) and the stakeholders? What are the specific interests and values ascribed to those positions and how can we make sure that they match the realities of those concerned? While it makes sense to differentiate between different interest groups and their level of involvement, the National Academies of Sciences, Engineering, and Medicine (NASEM) has problematized the arbitrary nature of grouping people into categories such as "communities", "stakeholders" and "public" (2016). The question remains crucial since the purpose of participation is dependent on a well-defined target group and the analysis of their needs, fears and hopes in order to find effective instruments of activation—especially since the perception of synthetic biology varies among different groups and contexts (see e.g. conservation and genetically modified plants vs. synthetic products such as vanillin or synthetically produced vaccines and drugs). In this respect, public attitude research shows that the perceptions of the field display a broad range of hopes and concerns for science and the acceptance of novel and potentially innovative methods is often conditional (Bhattachary et al. 2010; Leopoldina 2015).

The scope and designated sphere of deliberation (code of conduct, public participation, risk assessment, training and education etc.), the definition and scale of actors involved, and the forms and standards of evaluation are often only vaguely determined. In addition, while many call for participation (usually in the form of public engagement), the purpose, implications and practical consequences in terms of the regulatory decision-making process and the ways bottom-up deliberation should find its way into effective governance strategies needs for more nuanced and empirically grounded answers. With the advent of the upstream public engagement approach "informative" and "deliberative" participation often come across as two opposing poles, where the first is increasingly demoted as too expert-oriented and ignorant towards the valuable knowledge of the different "publics" and the latter has become synonymous with a more inclusive and democratic approach, they rather present different degrees of activation, which might be equally valid depending on their purpose within a specific project design.

Finally, in most policy papers and reports researchers or/and research councils are held responsible for initiating and maintaining a transparent and open dialogue with the public. The exclusive emphasis on either institutional or individual responsibilities seems short sighted and needs further exploration and concretization concerning the specific expertise required to initiate and maintain effective participation. What structural innovations are necessary to enable and facilitate said dialogue and what are the challenges and potential future pathways with regard to the roles, values and interests involved?

Given the frequent calls for more and better stakeholder and public involvement in reports on synthetic biology, we take cues from the literature on deliberative democracy and public participation in policymaking to illustrate what formats of participation might be conducive to meeting the demands for anticipatory governance.

References

Anderson J et al (2012) Engineering and ethical perspectives in synthetic biology. Rigorous, robust and predictable designs, public engagement and a modern ethical framework are vital to the continued success of synthetic biology. EMBO Rep 13(7):584–90. https://doi.org/10.1038/embor.2012.81. PMID: 22699939; PMCID: PMC3389334.

Bhattachary D, Calitz JP, Hunter A (2010) Synthetic biology dialogue report. Biotechnology and Biological Sciences Research Council and Engineering and Physical Sciences Research Council. Retrieved from https://bbsrc.ukri.org/documents/1006-synthetic-biology-dialogue-pdf/

Cummings CL, Kuzma J (2017) Societal risk evaluation scheme (SRES): scenario-based multi-criteria evaluation of synthetic biology applications. PLoS ONE 12(1):e0168564. https://doi.org/10.1371/journal.pone.0168564

Deutsche Forschungsgemeinschaft (DFG), Deutsche Akademie der Technikwissenschaften – acatech, Deutsche Akademie der Naturforscher Leopoldina – Nationale Akademie der

Wissenschaften (2009) Synthetische Biologie. Wiley-VCH. Retrieved from http://www.leopol
   dina.org/uploads/tx_leopublication/2009_NatEmpf_synthetische_biologie-DE.pdf
Deutsche Akademie der Naturforscher Leopoldina – Nationale Akademie der Wissenschaften
   (2015) Die Synthetische Biologie in der öffentlichen Meinungsbildung. Diskussion Nr. 3.
   Leopoldina. Retrieved from https://www.leopoldina.org/uploads/tx_leopublication/2015_Synt
   hetische_Biologie_DE_01.pdf
Douglas CMW, Stemerding D (2014) Challenges for the European governance of synthetic biology
   for human health. Life Sci Soc Policy 10:6. Retrieved from http://www.lsspjournal.com/content/
   10/1/6
ETC Group, Friends of the Earth, CTA (2012) The principles of oversight for synthetic biology. ETC
   Group. Retrieved from https://www.etcgroup.org/sites/www.etcgroup.org/files/The%20Princip
   les%20for%20the%20Oversight%20of%20Synthetic%20Biology%20FINAL.pdf
European Academies Science Advisory Council (EASAC) (2010) Realising European potential in synthetic biology: scientific opportunities and good governance. EASAC. Retrieved
   from   https://www.leopoldina.org/uploads/tx_leopublication/2010_EASAC_Statement_Synthe
   tic_Biology_ENGL.pdf
European Commission (2010) Europeans and biotechnology in 2010. Winds of change? Publica-
   tions Office. Retrieved from https://ec.europa.eu/commfrontoffice/publicopinion/archives/ebs/
   ebs_341_winds_en.pdf
International Risk Governance Council (IRGC) (2010) Guidelines for the appropriate risk governance of synthetic biology. IRGC. Retrieved from https://irgc.org/wp-content/uploads/2018/09/
   irgc_SB_final_07jan_web.pdf
International Union for Conservation of Nature (IUCN) (2019) Genetic frontiers for conservation: an assessment of synthetic biology and biodiversity conservation. Synthesis and key
   messages. IUCN. Retrieved from https://www.iucn.org/resources/publication/genetic-frontiers-
   conservationsynthesis-and-key-messages
Kelle A (2009) Ensuring the security of synthetic biology—towards a 5P governance strategy. Syst
   Synth Biol 3(1–4):85–90. https://doi.org/10.1007/s11693-009-9041-8
Kelle A (2013) Beyond patchwork precaution in the dual-use governance of synthetic biology. Sci
   Eng Ethics 19(3):1121–1139. https://doi.org/10.1007/s11948-012-9365-8
Köchy K, Hümpel A (eds) (2012) Synthetische Biologie: entwicklung einer neuen Inge-
   nieurbiologie? Themenband der interdisziplinären Arbeitsgruppe Gentechnologiebericht.
   Forschungsberichte der interdisziplinären Arbeitsgruppen. Berlin-Brandenburgischen Akademie
   der Wissenschaften (BBAW). Wissenschaftlicher Verlag. Retrieved from https://www.gentechno
   logiebericht.de/fileadmin/user_upload/Webseitendateien/Publikationen/Synthetische_Biologie_
   2012.pdf
Presidential Commission for the Study of Bioethical Issues (PCSBI) (2010) New directions: the
   ethics of synthetic biology and emerging technologies. Retrieved from https://bioethicsarchive.
   georgetown.edu/pcsbi/sites/default/files/PCSBI-Synthetic-Biology-Report-12.16.10_0.pdf
Stirling A (2015) Towards innovation democracy? Participation, responsibility and precaution in
   the politics of science and technology. STEPS Working Paper 78. STEPS Centre. Retrieved from
   http://sro.sussex.ac.uk/id/eprint/39718/1/Innovation-Democracy.pdf
Stirling A, Hayes KR, Delborne J (2018) Towards inclusive social appraisal: risk, participation
   and democracy in governance of synthetic biology. BMC Proc 12(Suppl 8):15. https://doi.org/
   10.1186/s12919-018-0111-3
SYNENERGENE (2017) Workshop report mutual learning in synthetic biology: policy Implications of RRI. Retrieved from https://www.synenergene.eu/sites/default/files/uploads/SYNENE
   RGENE_workshop_report2425042017finalsummary_0.pdf
Tait J (2009) Upstream engagement and the governance of science: the shadow of the genetically
   modified crops experience in Europe. EMBO Rep 10:S18–S22
The National Academies of Sciences, Engineering, and Medicine (NASEM) (2016) Gene drives on
   the horizon: advancing science, navigating uncertainty, and aligning research with public values.

The National Academies Press. Retrieved from https://nap.nationalacademies.org/catalog/23405/gene-drives-on-the-horizon-advancing-science-navigating-uncertainty-and

The Nuffield Council on Bioethics (2012) Emerging biotechnologies: technology, choice and the public good. Nuffield Council. Retrieved from https://www.nuffieldbioethics.org/publications/emerging-biotechnologies

The Royal Academy of Engineering (2009a) Synthetic biology: scope, applications, and implications. Retrieved from https://raeng.org.uk/media/fvwdlqmx/synthetic_biology.pdf

The Royal Academy of Engineering (2009b) Synthetic biology: public dialogue on synthetic biology. Retrieved from www.raeng.org.uk/news/publications/list/reports/Syn_bio_dialogue_report.pdf

The Royal Academy of Engineering (2011) Emerging biotechnologies. Nuffield council on bioethics response from the royal academy of engineering. Retrieved from https://raeng.org.uk/media/tbhhna4d/emerging_biotechnologies-2011.pdf

Trump BD (2017) Synthetic biology regulation and governance: lessons from TAPIC for the United States, European Union, and Singapore. Health Policy 121(11):1139–1146

Wagner S, Bollaert C (2013) Biosafety, biosecurity, and bioethics governance in synthetic biology: the "7P" approach. Appl Biosaf 18(4):178–186

Wallach W, Saner M, Marchant G (2018) Beyond cost-benefit analysis in the governance of synthetic biology. Governance of emerging technologies: Aligning policy analysis with the public's values. Hastings Cent Rep 48(1):S70–S77. https://doi.org/10.1002/hast.822

Zhang JY, Marris C, Rose N (2011) The transnational governance of synthetic biology: Scientific uncertainty, cross-borderness and the 'art' of governance. BIOS Working Paper No. 4. BIOS, London School of Economics and Political Science. Retrieved from http://royalsociety.org/uploadedFiles/Royal_Society_Content/policy/publications/2011/4294977685.pdf

Chapter 5
Governance and Public Participation

Abstract The chapter discusses different objectives of public participation in synthetic biology and suggests six focus areas across all governance dimensions, hard law, soft law, education, research impact, and research infrastructure, to design effective participation in synthetic biology. The authors identified three focus areas in participation—communication, representation, and evaluation—that are of particular importance for ensuring trust and justice as core values of SB governance. In addition, three intersecting focus areas in governance provide incentives for decision-making: resources, training, and data management. Furthermore, the authors highlight the question of responsibility as an area that requires special attention.

Keywords Objectives of Participation · Focus Areas, Communication · Representation, Evaluation · Resources, Training · Data Management · Responsibility · Formats

As outlined in the review of the literature and the policy reports on the governance of synthetic biology, calls for deliberation and public participation are rarely followed up with an elaboration of what deliberation and public participation would or could look like in practice. Moreover, this literature includes a strong focus on deliberation without distinguishing deliberation from other forms of participatory practices in science and policy. These observations are not dissimilar to critiques levelled against proponents of deliberative democracy, namely that the theory of deliberative democracy is often not matched by a reflection of its practical implications (e.g. Newell 2010). This, however, is not to say that there have not been significant efforts to improve public participation in emerging biotechnologies. Einsiedel (2012) provides an overview of, as she calls them, a number of experiments with public deliberative processes in several countries in the 1970s and 1980s following the first discoveries of the possibility of transfer of genes across species through scientific intervention. She shows, however, that these experiments were largely restricted to one format of deliberation: consensus conferences. While consultations on biotechnologies and, as some (e.g. Newell 2010) label it, efforts at democratizing biotechnologies are ongoing, the practical implications of improving public participation and deliberation are often mentioned as a sidenote, if at all. If, as the literature suggests, participation and deliberation are key to developing more effective and

S. Siewert et al., *Synthetic Biology and the Question of Public Participation*,
SpringerBriefs in Philosophy, https://doi.org/10.1007/978-3-031-16004-2_5

anticipatory governance structures in synthetic biology, then we need to consider what that means with regard to infrastructure, training, resources and organization. A failure to do so runs the risk of resulting in few, ineffective, tokenistic or ambiguous modes of participation that neither help the perception of synthetic biology at the science-society interface nor the effective governance of the field. Drawing on insights from the literature on public participation in policymaking, this section offers an exploration of what a more meaningful approach to public participation in the governance of synthetic biology could look like.

Public participation in governance and policymaking mainly refers to participation as political participation, that is "taking part in the processes of formulation, passage and implementation of public policies" (Parry et al. 1992, 16). Bishop and Davis define policy participation as "[…] a mechanism deployed by politicians and officials to expand those voices heard in the decision process" (2002, 26). According to Bishop and Davis, definitions of public participation usually include some, if not all, of the following features: (a) A measure of citizen involvement in decisions otherwise in the prerogative of governments; (b) A commitment to seeking views of those affected by a decision; (c) A transfer of authority from governments to citizens; (d) A transparent process which ensures citizens are informed about policy processes and decisions. The rationales for public participation are varied, but often rooted in what is labelled the democratic deficit, that is an apparent disconnect between the governing elites and the citizens in liberal democracies. As discussed in the section on the roots and definitions of governance, this rationale leads to calls for more deliberative forms of governance coupled with the view that policy issues, in science as well as other fields, are increasing in complexity.

This is important to note against the background of calls for more deliberation in the governance of synthetic biology, for while deliberation is a form of public participation, not all public participation is a form of deliberation. Deliberation is a mode of participation useful in specific instances, usually with a pre-defined purpose or goal: "Deliberation refers either to a particular sort of discussion—one that involves the careful and serious weighing of reasons for and against some proposition—or to an interior process by which an individual weighs reasons for and against courses of action" (Fearon 1998, 63). While deliberation is a powerful technique at the disposal of scientists and policymakers who are interested in opening avenues for critical public discussions, it is best employed for questions or issues that require careful deliberations, such as difficult moral or ethical issues in synthetic biology. As Rowe and Frewer (2004) point out, one also has to take into account the difference between public engagement activities that are geared towards policy decisions and those aimed at fostering the discourse about science and technology (emphasis on informal spaces of exchange). Designing participatory activities therefore raises the need to be specific about, and reflect upon, one's goals and objectives in involving the public. Here we would also argue that there is no predefined restriction on *when* to organize public activities (as advised by the Royal Society in 2004 with respect to nanoscience), but rather, one has to look for thresholds, where public activation would be productive.

For participatory efforts to be meaningful, the desired outcome, that is the intended influence of the public rather than the content of the outcome, needs to be specified. Here, the questions that policymakers, scientists and regulators need to think about when developing formats for participation are numerous. First, one needs to ask what goal participation in synthetic biology should serve. When involving the public in research activities, is the goal to change or adapt the activities or merely to inform the public about the developments in the field? When involving the public in regulation and policymaking activities, is the objective to change and adapt regulations or codes of conduct or is the objective to say that one has consulted the public, even if the end result remains unchanged? Is the goal to find out how the public view the societal value of new emerging biotechnologies such as synthetic biology in order to design research projects that reflect this view? In what follows, we suggest different objectives of public participation in synthetic biology. This list should not be read as an exhaustive list, but as an illustration of the varied formats that public input in synthetic biology could have. In the governance of synthetic biology, public participation could serve the following purposes (Fig. 5.1).

The list suggests that public participation in synthetic biology can happen at different levels of governance. As Bishop and Davis (2002) say, finding the appropriate mode of participation depends on the goals and purposes of participation. In the case of synthetic biology and other emerging biotechnologies, this implies that the purposes of participation are likely to be different at different levels of governance.

5.1 Focus Areas

In the following we identify six focus areas across all governance dimensions, hard law (national and international regulations and laws), soft law (codes of conduct, eg. by professional associations or research laboratories), education (training and skills for scientists, researchers and students), research impact (i.e. the outcomes /products emerging from research), and research infrastructure (funding structures, public/private research & development). Taken together, the elements represent the policy and regulatory spheres in which synthetic biology is governed as a scientific and biotechnical field.

We identified three focus areas in participation—communication, representation, and evaluation—that are of particular importance for ensuring trust and justice as core values of SB governance. In addition, we emphasize three intersecting focus areas in governance, which are decisive for future endeavors: resources, training, and data management. Taken together, we suggest that these six areas represent dimensions of public participation and governance that need to be considered and reflected upon by policymakers and those intending to organize participatory activities in order to ensure meaningful participation and deliberation in the field of synthetic biology.

Fig. 5.1 Objectives of public participation in synthetic biology

5.1.1   Focus Areas in Participation

5.1.1.1   Communication

When we consider aspects of communication in emerging technologies, the focus is usually on knowledge transfer and the presentation of research output to the public. Yet, communication is multifaceted and also pertains to the purpose of a participatory activity, the framework (scale, topic, timing), and the existing communication infrastructure. The latter also includes the communicative processes between those planning a participatory exercise and the public(s), industry, legislators, as well as the respective media channels. Last but not least, communication also relates to monitoring and evaluation in the sense that transparency about the impacts of participation requires open communication.

Science depends on international exchange, research is not bound to the national context and should not be limited to certain social groups. In order to reach different communities, and to communicate in a context-specific and culturally sensitive way, investment in better communication plans should be an ongoing process. This process requires long-term communicative strategies and goals, and it demands better training in science communication, for example in SB Masters and PhD programmes.

5.1.1.2 Representation

As many have pointed out (e.g. Abelson et al. 2007; Mitton et al. 2009; Weale et al. 2016), it is often not clear *who* the public is when policymakers, regulators or researchers talk about public participation. For obvious practical reasons, the term public cannot mean the whole population of a given country, so public participation activities require the selection of a sample of individuals from a given population. There are multiple questions the people in charge of organizing public participation have to ask: Is the goal to gain an insight on public views understood as a sample from a given population? Is the goal to involve the people who are most likely to be affected by innovations in synthetic biology, that is patients and consumers for example? If the latter, who does one turn to, individuals or interest and advocacy groups who represent the voices of those affected? In any mode of participation, how can one ensure that views and voices are accurately represented, and not dominated by special interests?

These are complex questions, the answers to which will most likely never be definite, but the engagement with it will contribute to building reflective public participation formats in the governance of synthetic biology. Once again, the answers are likely to depend on the purpose and the nature of a given participatory exercise. For some participation types and objectives, it might be useful to engage with interest and advocacy groups, for others one might wish to bring together people from different walks of life to gather input on questions of broad societal relevance. A careful consideration of whom to include in one's participatory efforts will also influence the evaluation of these efforts. In other words, determining the objective of participation will usually reveal the participants one wants to reach, which in turn affects the evaluation of the outcome and process of participation against one's original objectives.

Awareness for the needs of different groups is not only a question of access and representation in participatory activities, but also refers to addressing the provision of safe spaces, where people can voice their concerns. These issues tie back to other focus areas and trigger a range of interrelated questions: Are minorities adequately represented and involved? What have been successful strategies in creating and maintaining diversity in public engagement activities, and how can they be implemented across different projects in SB?

At the same time there are also differences in the question of who initiates the dialogue, for example organized civic society groups, state-initiated research programs, global independent research councils, private–public partnership projects? What are the interests, needs and expectations of those in charge? Public engagement even today is primarily in the hands of experts but when talking about representation the different forms of expertise and the balance and mediation between lay and expert knowledge becomes extremely important.

5.1.1.3 Evaluation

The need for specificity in one's objectives raises another important topic, that is the question of how to evaluate whether participatory activities have been successful or effective. The literature on the evaluation of public participation (e.g. Abelson and Gauvin 2006; Rowe and Frewer 2000, 2004) points to a paucity of empirical studies evaluating the success and effectiveness of public participation, which is explained by highlighting the challenges that emerge when seeking to evaluate public participation activities. According to Rowe and Frewer (2000) such challenges include the lack of agreement on a set of criteria against which to evaluate participation efforts. Moreover, what counts as successful or effective or, to use another set of evaluation terms, 'good' or 'bad' participation is context-specific and dependent on careful reflection on the objectives of public participation alluded to in the previous paragraphs.

To address this so-called evaluation gap, Rowe and Frewer (2000) propose an evaluation framework for public participation. The framework is divided into acceptance criteria—"related to the effective construction and implementation of a procedure" (Rowe and Frewer 2000, 11)—and process criteria that refers to the acceptance of a procedure. The acceptance criteria include principles such as representativeness, independence and influence, the latter referring to the ideal that any output of a given process should have a genuine impact on policy or, for our purposes, on research practices and regulations.

This gives rise to another important criterion, namely transparency about how public input was used, and what effect it had on the ultimate decision taken. This is where the areas of evaluation and communication are interlinked. Rowe's and Frewer's (2000) process criteria include task definition (being clear about the nature and scope of a given participatory exercise), resource accessibility (providing participants with the resources needed to fulfil their participatory roles), a form of structured or formalized decision-making (e.g. a specific format of participation such as a focus group) and cost effectiveness. While the individual evaluation criteria have been adjusted in various ways, the distinction between 'outcome' criteria and 'process' criteria appears as a common thread in the literature on the evaluation of public participation, which remains largely theoretical rather than empirical (Abelson et al. 2003; Abelson and Gauvin 2006). Evaluating public participation through the lens of outcome and process criteria underlines that public participation is not an end in itself, but that it offers a menu of possible modes and formats (processes) that aim to achieve an end (outcome). The effectiveness of public involvement needs further research and empirical data (Rudenko et al. 2018) and the acknowledgement of critical voices such as Wolt and Wolf (2018), who have cautioned against certain downsides of public comment when it comes to the approval of SB products.

"Lessons learned" should be made accessible for future SB projects. A central and collectively maintained and updated platform for participation in SB projects can facilitate the exchange of experiences on how to plan and install participatory elements. Such an infrastructure would help to minimize costly and time-consuming trial-and-error phases, and allows for a more sustainable testing of different methods. The actual implementation and maintenance (also with regard to data security) should not be project-based but rather thought of as a permanent structure that needs a place within EU or global science counseling.

5.1.2 Focus Areas in Governance

5.1.2.1 Resources

There seems to be an implicit assumption that the responsibility for coordinating and organizing public participation activities in synthetic biology lies with the research community. This implicit assumption raises questions regarding the infrastructure of research communities that are not adequately addressed. Organizing meaningful participation requires skills and experience, time and financial resources. Especially costs are a significant barrier for researchers, sponsors and participants (Stilring et al. 2018, 48–49). Deliberative formats of participation require extensive time for preparation, and frequently involve the hiring of external moderators to support the facilitation of citizens' conferences for example. Issues of synthetic biology pose challenges in this regard. Due to their complex technical nature, the translation of production processes and modes of action "require a high degree of training or a significant amount of editorial effort to demystify the language" (Stirling et al. 2018, 48). In the end SB public engagement might occasion higher costs in designing appropriate and effective modes of interaction while the value of shared perspectives and knowledge is difficult to measure. Costs and benefits should be balanced—keeping in mind that achievements might be hard to quantify and yet every genuine effort holds the potential for relationships, trust, and insight. In the long run, willingness to invest will be an important factor of credibility for everyone involved (Stirling et al. 2018).

This also applies on the level of research funding.[1] Resources for staff training and for the organization of public participation should be included in funding applications. This has been the case for many major projects. Especially synthetic biology

---

[1] In terms of funding, synthetic biology has been the subject of long term governmental efforts. The US and the UK in particular provided early road maps for the development of the field. The European Commission (EC) initiative on New and Emerging Technologies (NEST), the Horizon 2020 and the national funding agencies played a vital role for the advancement of SB research (Donati et al. 2022).

has seen a fair amount of "public acceptance" research (see e.g. Leopoldina 2015). As Marris and Rose argue: "For reasons that may be genuine or symbolic, 'public engagement' has become almost obligatory in major programmes of publicly funded research aimed at the development of novel biotechnologies" (2010). There are numerous inspiring initiatives of public participation and institutions promoting public engagement activities in biotechnology. The EU Horizon 2020 on Public Engagement in Responsible Research and Innovation, ELSI and the Human Genome Project as well as other programmes have been at the forefront in establishing standards and in defining ethical guidelines.

However, compared to other funding areas, societal and ethical aspects of SB research still make up only a small percentage. In a 2010 survey on the trends in synthetic biology research funding in the US and Europe, the Woodrow Wilson International Center for Scholars concluded that since 2005 the U.S. government had invested approximately $430 million on research related to synthetic biology, but only 4% of the money was used to examine the ethical, legal and social implications of synthetic biology. The European Union and three European countries had spent approximately $160 million on synthetic biology research with around 2% going toward implications research (Woodrow Wilson Center 2010). At the same time, the funding allocated to societal aspects of SB subsumes a number of different disciplines, methods and topics. Research on or through participation is just one aspect among many.

In fact, what seems missing is a long-term approach beyond specific funding lines, serious efforts to turn the lessons learned into funding strategies, and an awareness concerning the changing expectations and demands posed on researchers when participation is taken seriously as an operating mode of governance.

These are merely examples of the practical questions that need to be addressed with regard to research infrastructure. At the very minimum, there needs to be transparency about what resources, financial and otherwise, are required to meet participatory aims in synthetic biology.

5.1.2.2 Training

How can we promote motivation and skill for public engagement on the part of researchers, research councils, and policy representatives to create attention and inspire enthusiasm in the public(s)? To ask what scientists can do for the acknowledgement of SB research first and foremost requires a self-reflective process and an understanding of how the publics are perceived and understood in different scientific contexts (Hansen 2010). It is important to provide orientation throughout this process. The interdisciplinary public dialogue organized by the BBSRC and EPSRC offers a set of guiding questions that researchers working in the field might consider at each and every step of their endeavor: "What is the purpose?"; "Why do you want to do it?"; "What are you going to gain from it?", "What else is it going to do?"; and "How do you know you are right?" (Bhattachary et al. 2010). At times, researchers might find it hard to leave their comfort zone and to make their findings accessible within established frames of interpretation—not least for ethical reasons (Dudo and Besley 2016). The American Association for the Advancement of Science (AAAS) has emphasized that a focus on responsiveness, resonance with values, and mutual learning arc not common within much of the science community. Here it is crucial to develop "a new understanding of the intent and ethics of framing as a communication strategy, in a way that resonates with a scientific worldview" (2016).

Concomitantly it requires adequate time and resources to explore ethical questions and wicked problems of SB, while at the same time researchers need to be well versed in the methods and practical implications of PE. Research and funding institutions need to find ways to accommodate public engagement education into research agendas and training programmes, without further pressure on scientists to integrate an additional demand. This also affects institutional cultures and the willingness to reach out and alter standards of scientific gratification (and certification). Last but not least, there are of course national differences and expectations regarding the societal role and function of scientists that have to be taken into account.

5.1.2.3 Data Management

Digitization and digital culture bring new challenges with respect to the activation and recruitment of different social groups for participatory activities. While access and data management play a major role, there is also the aspect of digital literacy. At the same time, digital participation might offer additional opportunities to address audiences that are often hard to reach through more traditional recruitment processes. This requires a certain openness with regard to new media channels and formats, and an expertise on how to employ media communication. Ethical dilemmas connected to aspects of empowerment and corporate hosting (e.g. the use of commercial conference tools) have to be considered throughout the entire process of communication. Practitioners also have to be aware that the methodological landscape, and with it established conventions and standards, might alter. Will there be certain formats and activities that prove to be more suitable for the digital realm, and how does that affect the range of possibilities? Will online surveys and consultations prevail, simply because they seem easier to handle with respect to data management and the evaluation of results? What happens to the informal in-between spaces at citizen conferences and forums? How can they be transferred into the digital arena?

The same holds true for the visibility of certain topics. Seemingly less prominent issues in synthetic biology might be neglected due to prevailing healthcare issues, such as the COVID-19 pandemic. Those responsible for public activation will have to be innovative to gain traction for their involvement strategies. Finally, questions of representation do not lose their significance. It remains to be seen if digital participation will be able to assemble representative positions here, as the digital communication spaces are prone to discrimination of certain socio-economic groups.

5.2   Responsibility

In addition to the above focus areas, we highlight the question of responsibility as an area that requires further attention. While the question of who the stakeholders are, has been a topic of reflection in a number of policy reports, a careful assessment of who is responsible for ensuring meaningful public participation is often neglected. It is important to look at the different interests, fears and hopes of those involved, and the ways they are connected to individual and institutionalized moral, legal and social responsibilities. In practice, the criteria and standards required for the attribution of responsibility depend on the specific circumstances and existing framework conditions under which people carry out their actions. Intention and intent, skills and knowledge, tasks and roles, the work environment and legal regulations are factors that must be taken into account when attributing responsibility (Heidbrink 2003, 30ff.).

We draw a distinction here between responsibility and accountability. Accountability usually has legal and political connotations, at least in liberal democracies. To be accountable for an action, an outcome or a policy means that a person or legal entity can be made answerable for her actions and the resulting consequences. Appeals procedures or parliamentary investigative committees are examples of institutions that serve the goals of finding out who or what is accountable for a specific outcome if something goes wrong. The distinction between accountability and responsibility emerges in the sense that one can be accountable for an action or an outcome without having been responsible for it.

In relation to participation in the governance of synthetic biology, the above thoughts demand a nuanced reflection about who is responsible for carrying out participation activities. We refer to responsibility here as the act of organizing and conducting participatory exercises, which includes the responsibility of ensuring that outcomes of these exercises are adequately considered in future research. Researchers, research funders and stakeholders all hold responsibilities with regard to the improvement of public participation in emerging biotechnology research. In the next paragraphs we elaborate on these responsibilities. In line with our concerns for the practical implications of calling for more public participation in synthetic biology, our focus here is on responsibility rather than accountability. The discussion on responsibility shines a light on some of the practical challenges of improving public participation, whilst a discussion on accountability would lead us into the realm of legal and public duties, which is beyond the scope of our report. Just like in the political arena, the individuals and entities responsible for the organization of participatory activities can be the same, or distinct from, the individual and entities accountable for how the outcomes of such activities are integrated into the research process.

5.2.1 Researchers

In policy reports on synthetic biology, it is often implied that the people responsible for carrying out public participation activities are the researchers. The roles, tasks and public images of scientists, however, vary according to discipline, institution, state or country. The heterogeneity of expectations and interests has to be taken into account. The assessment and assignment of responsibilities should begin with a close inspection of the actual situation of scientists in their particular working environment. Many researchers understand their task in delivering valid results that might improve the situation of certain (patient) groups or consumer groups without efforts in designing participatory activities. According to their understanding, time is better spent in furthering research. In addition, researchers, who have never had experiences in designing and implementing public engagement activities, will likely feel wary about their role in carrying out such activities.

Moreover, ambitions of public participation frequently collide with the realities of the research process. Scientists have to raise research funds, are expected to publish their results, teach, and engage in national and international collaborations. There is often not enough time for participatory projects and time-consuming deliberative processes, even if, from a research ethics perspective, it would be sensible to do so against the background of the issue characteristics of synthetic biology highlighted in Chap. 4. Institutional commitments that pertain to regulatory considerations in a field can make it difficult for researchers to know when and where to share their ideas. Gene technology, for example, is an innovation-driven research field where patenting is a complicated, but highly desired and defining, feature of scientific processes, meaning the sharing of knowledge, ideas and results may not always be possible under the constraints of patent laws. To present work in progress to a broader audience might come at a price in such a highly competitive field of private and public interest groups. Furthermore, funding structures and predefined career paths often do not allow additional (often time consuming) involvement in participatory activities or the development of participatory projects. Structural improvements that would grant researchers the time and resources to take part, learn and develop their own project-specific objectives and participation designs, supported by hosting institutions and regulatory bodies, might foster tighter bonds between science and society. In other words, if research institutions, society and policymakers expect researchers to engage with the public about their research, there needs to be an open discussion about the requirements in relation to infrastructure, skills and incentives to carry out meaningful participation activities. Researchers should not be blamed for a lack of participatory efforts, if this lack of participation is rooted in legal, regulatory or infrastructure constraints.

5.2.2 *Research Funders*

Research funders set standards for scientific quality, they connect scientists, enable scientific progress, they provide funding, and promote innovation. They can play a pivotal role in empowering scientists to invest time and resources in deliberative processes, public consultations and other formats. They have a responsibility for improving public participation by providing researchers with extra funds for such activities. Some funders already do this by offering integrated or separate funding streams for public participation. Some funders, however, require research teams to develop elaborate plans for public participation without offering additional funds, or even an acknowledgement that such activities will add time onto the length of a research endeavour. Additionally, some funders require public participation plans without a reflection of the specific purpose that public participation can serve in each instance. This comes back to our earlier discussions about the purposes and types of participation. Here, research funders not only have a responsibility to provide adequate funding and skills training in public participation, but also to reflect and deliberate on the goals of public participation in every application they assess in order to ensure that participation can be meaningful.

5.2.3 *Stakeholders*

Stakeholders, ranging from industry partners to civil society organizations and patient groups, might have differing, even competing, perspectives on engagement strategies and the role of the public(s) in the decision-making process. While industry partners or private research institutes might support participation as a form of horizon scanning and an authentic marketing asset in promoting synthetic products, they can also be hesitant when it comes to deliberative projects that might potentially prolong approval and introduction of said products. Civil society organizations have a distinct interest in involving the public and to strengthen the legal rights and terms of participation of the social groups concerned—as e.g. in the case of synthetic crops and farming. However, their calls should be backed by trustworthy evaluation of the needs and wants of those affected. All stakeholders have the responsibility to be transparent about their objectives in participatory endeavors, and to work with scientists and other experts in the field, for example by sending representatives to public and deliberative forums and/or to share the knowledge, experience and attitudes of the groups they represent within these forums.

5.2.4 Legislators and Regulators

In an ideal scenario of policymaking that acknowledges the importance of evidence in informing decision-making, legislators consider scientific information and risk assessments as a foundation for regulatory measures. This includes an awareness of the complex ethical dilemmas that emerge in the fields of some emerging biotechnologies, and the question of how to communicate them. Public engagement can help legislators to get a comprehensive picture of the potential harms, opportunities, and ethical concerns that synthetic biology entails. By fostering and enabling structural change and opportunities for scientists to invest time and resources into participatory activities, legislators would meet their responsibility to ensure that researchers are able to identify the wicked problems of synthetic biology in dialogue with the public. It is the responsibility of legislators and regulators to provide opportunity structures under which participation can thrive. Already in 2010, the European Academies Science Advisory Council called upon public policy-makers at national and EU level, to consider "their responsibilities for policy issues associated with security, ethics and public dialogue—clarifying who else shares the responsibilities" (EASAC 2010, 23).

In sum, participation can be viewed as part of a prospective precautionary principle, where science and society (together) take the initiative, and hold responsibility, to further innovation and research whilst avoiding or mitigating against potential harms. There is not just one person, group or stakeholder responsible for meaningful participation, but rather different competences, knowledge and experiences need to come together in order to facilitate such participation.

The following Table 5.1 helps to identify steering effects and orientation points for future endeavors in the field. The columns "considerations", "future pathways", and "responsibility" emphasize social, ethical and organizational aspects that, when taken into account, might offer productive paths to action. The column "responsibility" lists authorities and positions that should/could be involved in participatory activities in the different focus areas of participation and governance.

The following illustration combines the dimensions of governance, objectives of public participation and the six focus areas. The overview is not prescriptive but acquires a heuristic function. The governance dimensions presented here—from hard law to aspects of research infrastructure—are interconnected. Certain considerations, e.g. with regard to communicative strategies, questions of representation and appropriate resources can be debated for all dimensions and call for a more integrated approach. At the same time the objectives of participation are equally dependent on the topic, purpose and target of engagement and are not linked to one dimension alone (Fig. 5.2).

Table 5.1 Focus areas in participation and governance

Focus areas participation	Considerations	Future pathways	Responsibility
Communication	• Purpose of public engagement (PE) • Framework of engagement activity (scale, topic, timing) • Form and mode of output • Communication infrastructure (mediation of communicative process between PE practitioners, public(s) and policy representatives) • Monitoring and evaluation	• Transparency regarding design, conduct, and dissemination of research project • Definition of purpose (policy-oriented, research-oriented, discourse-oriented?) • Transparency of impact of PE in decision-making processes • Involvement of communication experts to develop, implement and evaluate public exchange • Rules of engagement • Effective choice of communication channels/infrastructure • Monitoring and evaluation of PE activity impact on all levels of governance	• Legislators • Research councils • SB researchers • PE practitioners • Communication experts • Quality control
Representation	• Position of initiator • Selection and constellation of participants • Addressing of target group • Space of communication • Presentation of output	• Information about initiator and facilitator of dialogue (interests, needs and expectations) • Systematic consideration of the voices and needs involved • Open and transparent nomination process of representatives • Balance between different interest groups • Balance between expert and lay knowledge • Acknowledgement of different forms of expertise, values and beliefs • Open and safe spaces of negotiation • Continuous reflection and evaluation of ethical and social questions • Culture- and status sensitive communication of results (barrier-free)	• Legislators • Regulators • Research councils • SB researchers • PE practitioners • Communication experts (focus equal opportunities) • Civil society organizations
Evaluation	• Scale of research output • Evaluation standards • Sustainability of PE activity • Impact on policy	• Standardization of PE evaluation • Monitoring of PE activities • "Lessons learned" (open repository for public participation) • Transparent evaluation of policy impact	• Legislators • Regulators • SB researchers • PE practitioners • Communication experts

(continued)

Table 5.1 (continued)

Focus areas governance	Considerations	Future pathways	Responsibility
Resources	• Funding structures • Infrastructure and personnel • Cooperation	• Time and resources to experiment with different instruments of public participation • Experimental and error-friendly integration of participatory elements in funding guidelines • Seed funding for participatory-based projects (testing phase prior to submission of proposals) • Special funding status for participatory-based projects • New paths for private-public partnerships and interdisciplinary cooperation	• Legislators • National and international funding institutions and programmes • Research councils • Research institutions
Training	• Expertise at research institutions • Infrastructure and resources • Institutional cultures	• Adequate time and resources to explore ethical questions and wicked problems of SB (and biotechnologies in general) • Participation as part of curricula of SB training and study programmes • Training in the methods and practical implications of PE • Acknowledgement and awareness of necessity to enable and foster science-society dialogue as part of democratic cultures • Open spaces of negotiation • Continuous reflection and evaluation of ethical and social questions	• Legislators • Regulators • SB researchers • PE practitioners • IT + design experts • (Data science educators)
Data management	• Access to participation platforms • Digital literacy • Data protection	• Infrastructure and resources • Information and training in online-participation • Data sovereignty • Data protection management and monitoring	• Legislators • SB researchers • PE practitioners • Communication experts • Research institutions • State and federal research programmes • International science exchange programmes • Civil society organizations

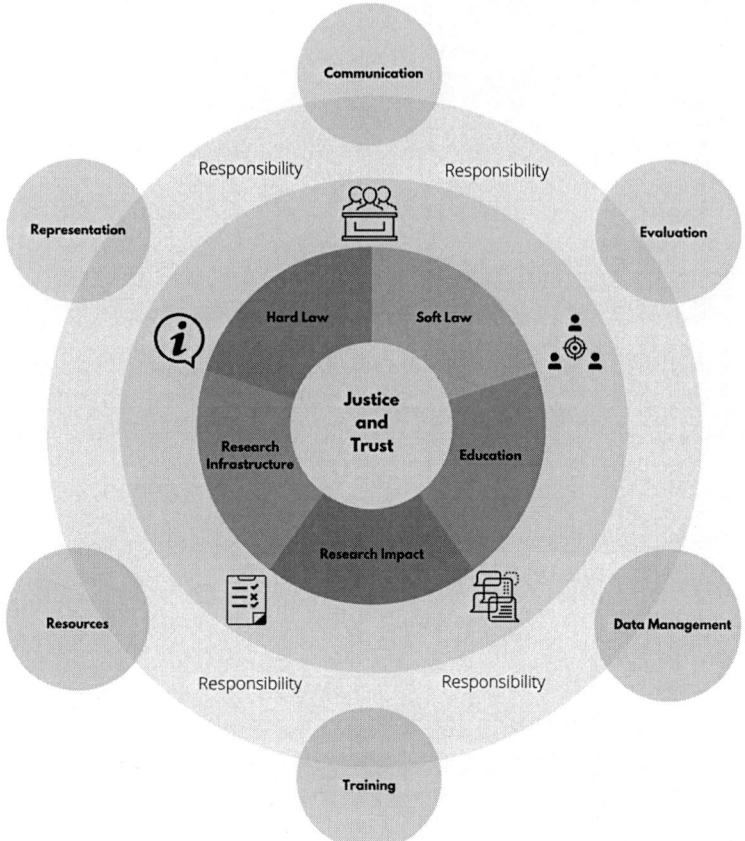

Fig. 5.2 Integrated approach of core ethical values, governance dimensions, objectives and focus areas

References

Abelson J et al (2003) Deliberations about deliberative methods: issues in the design and evaluation of public participation processes. Soc Sci Med 57(2):239–251

Abelson J, Gauvin FP (2006) Assessing the impacts of public participation: concepts, evidence and policy implications. In: Research Report P|06. Ontario: Canadian Policy Research Networks

Abelson J et al (2007) Bringing 'the public' into health technology assessment and coverage policy decisions: from principles to practice. Health Policy 82(1):37–50

American Association for the Advancement of Science (AAAS) (2016) Theory of change for public engagement with science. AAAS. Retrieved from https://www.aaas.org/sites/default/files/content_files/2016-09-15_PES_Theory-of-Change-for-Public-Engagement-with-Science_Final.pdf

Bhattachary D, Calitz, JP, Hunter A (2010) Synthetic biology dialogue report. Biotechnology and biological sciences research council and engineering and physical sciences research council. Retrieved from https://bbsrc.ukri.org/documents/1006-synthetic-biology-dialogue-pdf/

Bishop P, Davis G (2002) Mapping public participation in policy choices. Aust J Public Adm 61(1):14–29. https://doi.org/10.1111/1467-8500.00255

Deutsche Akademie der Naturforscher Leopoldina – Nationale Akademie der Wissenschaften (2015) Die Synthetische Biologie in der öffentlichen Meinungsbildung. Diskussion Nr. 3. Leopoldina. Retrieved from https://www.leopoldina.org/uploads/tx_leopublication/2015_Synthetische_Biologie_DE_01.pdf

Donati S et al (2022) Synthetic biology in Europe: current community landscape and future perspectives. Biotechnol Notes 3:54–61. https://doi.org/10.1016/j.biotno.2022.07.003

Dudo A, Besley JC (2016) Scientists' prioritization of communication objectives for public engagement. PLoS ONE 11(2):e0148867. https://doi.org/10.1371/journal.pone.0148867

Einsiedel E (2012) The landscape of public participation on biotechnology. In Weitze M-D et al (eds) Biotechnologie-Kommunikation. Kontroversen, Analysen, Aktivitäten. Acatech-Diskussion. Springer Vieweg, pp 379–412. Retrieved from https://www.acatech.de/wp-content/uploads/2018/03/acatech_DISKUSSION_Bio_Kom_WEB.pdf

European Academies Science Advisory Council (EASAC) (2010) Realising European potential in synthetic biology: scientific opportunities and good governance. EASAC. Retrieved from https://www.leopoldina.org/uploads/tx_leopublication/2010_EASAC_Statement_Synthetic_Biology_ENGL.pdf

European Commission (2020) EU horizon 2020 on public engagement in responsible research and innovation. Publication Office. Retrieved from https://ec.europa.eu/programmes/horizon2020/en/h2020-section/public-engagement-responsible-research-and-innovation

Fearon JD (1998) Deliberation as discussion. In: Elster J, Przeworski A (eds) Deliberative democracy. Cambridge University Press, pp 44–68

Hansen J (2010) Biotechnology and public engagement in Europe. Palgrave Macmillan

Heidbrink L (2003) Kritik der Verantwortung. Velbrück Verlag, Zu den Grenzen verantwortlichen Handelns

Marris C, Rose N (2010) Open engagement: exploring public participation in the biosciences. PLoS Biol 8(11):e1000549. https://doi.org/10.1371/journal.pbio.1000549

Mitton C et al (2009) Public participation in health care priority setting: a scoping review. Health Policy 91(3):219–228

Newell P (2010) Democratising biotechnology? Deliberation, participation and social regulation in a neo-liberal world. Rev Int Stud 36(2):471–491

Parry G, Moyser G, Day N (1992) Political participation and democracy in Britain. Cambridge University Press

Rowe G, Frewer LJ (2000) Public participation methods: a framework for evaluation. Sci Technol Human Values 25(1):3–29

Rowe G, Frewer LJ (2004) Evaluating public-participation exercises: a research agenda. Sci Technol Human Values 29(4):512–557

Rudenko L, Palmer MJ, Oye K (2018) Considerations for the governance of gene drive organisms. Pathog Glob Health 112(4):162–181. https://doi.org/10.1080/20477724.2018.1478776

Stirling A, Hayes KR, Delborne J (2018) Towards inclusive social appraisal: risk, participation and democracy in governance of synthetic biology. BMC Proc 12(Suppl 8):15. https://doi.org/10.1186/s12919-018-0111-3

Weale A et al (2016) Introduction: priority setting, equitable accesses and public involvement in health care. J Health Organ Manag 30(1):736–750

Wolt JD, Wolf C (2018) Policy and governance perspectives for regulation of genome edited crops in the United States. Front Plant Sci 9:1606. https://doi.org/10.3389/fpls.2018.01606

Woodrow Wilson International Center for Scholars (2010) Trends in synthetic biology research funding in the United States and Europe. Synthetic Biology Project. Research Brief. Retrieved from https://www.wilsoncenter.org/sites/default/files/media/documents/publication/final_synbio_funding_web2.pdf

Chapter 6
Case Study: COVID-19 Vaccine Development

Abstract When talking about public participation and the regulatory decision-making process, it is crucial to think about the right time and purpose to involve different stakeholders. The chapter explores circumstantial prerequisites for participatory approaches on the basis of a more recent case study: synthetically produced vaccines against the coronavirus SARS-CoV-2. The COVID-19 pandemic made synthetic biology a global issue—in a way that is unprecedented. Yet, its implementation also poses questions for participation for future vaccine and drug production: How can and should we engage the public on ethical questions of emerging and potentially disruptive technologies during a public health emergency? How do certain focus areas, particularly communication and resources, data management and training, provide valuable avenues in dealing with future emergency applications of SB?

Keywords Pandemic · COVID-19 · mRNA Vaccine · Synthetic Vaccines · Emergency Authorization · Communication · Resources · Training

The global efforts to find a vaccine against COVID-19 provide an interesting case study for how (not) to engage with the public. The state of emergency necessitated a deviation from any standard participatory processes that may have been in place to accelerate the search for a suitable vaccine. Not only that, but the very nature of the pandemic implied that participatory processes were not possible, at least not in the beginning of the pandemic and not in person. Despite these circumstances concerns for public involvement have become ever more pressing. The accelerated pace at which research into, and trials of, new therapeutic and prevention interventions happened, caused some members of the public to question the trustworthiness of outcomes regarding safety, efficacy, and ethics. To put it bluntly, the current frameworks for involving the public in research did not withstand the test of crisis. Efforts to involve the public(s) in deliberative processes need to be refined in order to make them more resistant to crisis situations. Data security and questions of access and representation (e.g. in terms of the composition of focus groups, citizens committees, consumer or patient groups) are under renewed scrutiny. Policymakers and healthcare decision-makers will likely have to invest in established and new forms of digital interaction to maintain dialogue with the public. When participation is restricted almost exclusively to the digital realm, we need to ask: What are the (structural) prerequisites to enable engagement for everyone? When vaccine development happens at "pandemic speed" (Lurie et al. 2020), could and should participation pick up the pace? Can it still work as an end-to-end process and potential operating mode of governance?

In September 2022, the WHO listed 172 COVID-19 vaccines in clinical development and 199 vaccines in pre-clinical development (WHO 2022). The types tested so far vary from live-attenuated vaccines, inactivated vaccines, subunit vaccines, virus-like particles (VLPs), viral vectors (replicating and non-replicating), DNA and RNA vaccines. However, 40 out of 172 candidates in development are based on RNA technology, making it the second most used platform after protein subunit vaccines.[1] In one of the most prominent examples, the German company BioNTech and its pharmaceutical collaborator Pfizer departed from more traditional methods by using a vaccine that contains parts of the genetic code of the virus. The vaccine carries a specific synthetically produced stretch of messenger RNA (ribonucleic acid), which then is injected into the human body. The cells "read" the mRNA as an instruction to build proteins, such as the spike-protein. Consequently, the body begins to fight the virus proteins made by our cells and produces antibodies. When Pfizer-BionTech submitted its vaccine for FDA approval in 2020, Phase 3 of clinical testing demonstrated a 95% efficacy in preventing a COVID-19 infection.[2]

Usually it takes 10 years to introduce a vaccine to the market since it requires a multi-stage clinical trial process: Preclinical Testing; Phase 1 (Safety Trials); Phase 2 (Expanded Trails); Phase 3 (Efficacy Trials); Regulatory Review and Approval. At the last stage regulators in each country review the trial results and decide whether to approve the vaccine or not. During a pandemic, a vaccine may receive emergency use authorization before getting formal approval (Thanh Le et al. 2020). BioNTech was not the only candidate working with synthetic components or designs. Moderna, CureVac and Inovio Pharmaceuticals built upon experimental vaccines containing lab-made strands of RNA and DNA. There is another prospective in synthetic biology to engineer a universal vaccine that in the future could be more potent and prove even more effective against a range of coronaviruses: the nanoparticle approach (Shin et al. 2020).

While the depth of engineering varies, from a sequence of the genetic code to the atomic level, synthetically produced material is delivered into the human body (even

---

[1] See the WHO's "COVID-19 Landscape of novel coronavirus candidate vaccine development worldwide", https://www.who.int/publications/m/item/draft-landscape-of-covid-19-candidate-vaccines.

[2] See press release of Pfizer, November 18, 2020: https://www.pfizer.com/news/press-release/press-release-detail/pfizer-and-biontech-conclude-phase-3-study-covid-19-vaccine. In light of the competitive spirit between different vaccine candidates new developments in the "race" for a vaccine were not surprising but caused irritation: BioNTech and Pfizer announced an efficacy rate of 95% shortly after Moderna released their findings which exceeded the number Pfizer had given earlier in November 2020. Pfizer's initial interim analysis showed a protection rate of 90%. The company readjusted the numbers just two days after Moderna had published an interim analysis with news of an even higher efficacy of 95%: https://investors.modernatx.com/news-releases/news-release-details/modernas-covid-19-vaccine-candidate-meets-its-primary-efficacy.

the cell) to trigger a certain response. The complex ethical questions connected to such research endeavors have to be addressed—just when and how?

When talking about public participation and the regulatory decision-making process, it is crucial to think about the right time and purpose to involve different stakeholders and the public(s). The appropriate tool of activation also depends on the severity of the problems and issues at hand. Synthetic technologies in cancer therapy or the treatment of the HIV-virus trigger different questions, they are also limited to certain groups of patients. The COVID-19-pandemic made synthetic biology a global issue—in a way that is unprecedented. Despite this, the discussions around, and sometimes even the opposition to the newly available vaccines focus on the novel make-up of the vaccines but hardly mention the SB technology behind it specifically (Paul et al. 2022). In general, "public information on the specific SARS-CoV-2 antigen(s) used in vaccine development is limited" (Thanh Le et al. 2020, 305).

But why is it important to raise awareness for the synthetic components of vaccines? What is the added value in enabling deliberative processes on the issue—especially when time is of essence as in the case of the COVID-19 vaccine development?

Attitudes towards vaccination vary greatly between countries. Vaccine acceptance depends on age, gender and social status (Lazarus et al. 2020). Compared to numbers in June 2020, public hesitancy towards a COVID-19 vaccine has increased globally. Less people would "definitely" get a vaccine primarily due to safety concerns connected to an accelerated vaccination development. Most alarmingly, hesitancy relates to a lack of trust in the government to make the right decision with respect to a new vaccine (Kantar 2020). There is no guarantee that participatory efforts will improve trust in institutions and vaccine recommendations (or even mandates), but not engaging the public at all, or only in a cursory way, will likely lead to a stalemate in the complex society-science relations in times of crises.

Trust is contextual, it is precarious and should not be taken for granted. In certain situations, participation might even endanger trust in medical and political authorities. An open deliberative process on the risks of synthetic components might jeopardize an already established pipeline and institutional backbone to solve a major public health crisis. At worst, it might further destroy trust in current processes for developing vaccine and other medical interventions. So one of the major goals of participation could be to strengthen the trustworthiness of national and international regulatory organizations such as the STIKO (Standing Committee on Vaccination), the ZBKS (The Central Committee on Biological Safety) or PEI (Paul-Ehrlich-Institut - Federal Institute for Vaccines and Biomedicines) in Germany or the EMA (European Medicines Agency) and WHO at a European and international level—especially in view of their common achievements and hard-earned agreements.

Having said this, why should we think about participatory governance as rewarding even when it involves risks, high stakes and enormous organizational effort? We argue that moments of crisis prove to be vital for long-term effects of participation as a basis for trust in synthetic biology and the emerging biotechnologies. Public activation helps to build a sustainable and authentic interest in establishing a trustworthy relationship between science, society and regulators (Siewert and Dabrock 2020). As synthetic vaccines have proven to be effective, public engagement can build upon the learnings and the success of mRNA therapies to discuss benefits and risks of future applications. To use a real-life event that people around the world have experienced and witnessed provides a more tangible approach to the often abstract and complex technologies of synthetic biology. The intent here is primarily geared towards building trust in the democratic process and decision-making in innovation-driven disciplines. It requires careful deliberation of when, and for what purpose participation is adequate.

With respect to vaccination, every threshold offers a potential opportunity of interaction. Each step in the process, from the exploratory to preclinical stage, the preclinical evaluation to testing, from testing to approval and possibly a change of legislation might be an apt moment to involve external expertise and feedback on the risks and benefits of vaccines based on synthetic material. In fact, there are a number of aspects that require critical attention. The shortening of clinical phases carries the risk that delayed side effects may only be detected after the vaccine has been widely used (Zylka-Menhorn and Grunert 2020). This in fact might be true for any other contagious disease in the future. And although the cost and amount of material needed to produce those new vaccines are relatively low compared to other methods of development, they require novel development and manufacturing platforms. The funding and future (mis)use of those technical infrastructures, however, receives only little attention. In this respect, synthetic vaccines relate to fundamental SB issues: uncertainty regarding outcome, new infrastructure, and product distribution.

Ethical questions have to be addressed at the beginning and throughout the whole process (Arvay 2020; Jiang 2020). As Jiang stated early on in the pandemic:

> Testing vaccines and medicines without taking the time to fully understand safety risks could bring unwarranted setbacks during the current pandemic, and into the future. The public's willingness to back quarantines and other public-health measures to slow spread tends to correlate with how much people trust the government's health advice. (2020, 321)

In addition, the way different governments handle(d) the COVID-19 pandemic—especially with respect to a globally fair vaccine allocation system—will be a "critical component of future pandemic preparedness" (Lurie et al. 2020, 1973). In fact, the COVID-19 Vaccines Global Access (COVAX) program has already seen a fair amount of critique (Taylor 2022).

The decision to involve the public depends on the objectives of participation and the specific contexts for which participation is designed. **To inform, consult, deliberate, gather and advise** at different stages might provide researchers, the public and legislators with a certain security in times of (heightened) uncertainty and risk. While the established safety and security protocols should remain in the hand of experts, a vaccine against a threat like the coronavirus is a concern of public welfare and should be handled as such.

To gather information is particularly relevant at the exploratory and preclinical state. What is being said about vaccine development in news media, public statements, social media, in advertising, policy documents or reports? This provides a picture of the current debates and helps identify key stakeholders and experts in the field. It also involves thoughts on how the information gathered might be evaluated and taken into account for the next steps in involving the public(s).

Consultations might prove to be productive on the acceptance of the synthetic design, delayed side effects, possible rapid mutation, (global) distribution of vaccine (bioeconomy) and also on novel vaccine development paradigms and the corresponding change in regulatory processes.

Deliberative modes might offer insights on the values and principles of an emergency authorization, especially at the preclinical state, but also with respect to a fair national and global allocation system.

Information is required primarily during the clinical stage and during review and approval processes. External communication should entail details on the candidates, the different vaccine types, national and global vaccine development, safety of novel development and manufacturing platforms, and the distribution of funding. In terms of licensing and the recommendations by the regulatory organizations such as the STIKO in Germany, broadly disseminated information on the approval process, planned global application and future modes of production could improve trust in accelerated procedures and the implementing institutions.

In addition, a detailed analysis of current science communication and the foundations of digital trustworthiness is particularly pressing in view of the current "infodemic" and widespread misinformation throughout the COVID-19 pandemic (see Nguyen and Catalan-Matamoros 2020).

While hard law pertains to laws and regulation, there are other governance dimensions, which are also closely related to the success of future legislation. How the effectiveness of vaccines is being communicated by different stakeholders is also essential. The introduction of the AstraZeneca vaccine in Germany serves as a good example to illustrate the power of communication in terms of public acceptance. AstraZeneca had been discredited early on due to inconsistent messages by regulators and the media. The recommendations by the Standing Committee on Vaccination in Germany (STIKO) had been more than confusing regarding the tolerability in elderly. While the STIKO had first cautioned against its use for people over 65, a revised recommendation issued shortly after saw it particularly suitable for elderly people. The first message, the supposedly ineffectiveness in the elderly, circulated in the media, initiated by the German Handelsblatt in February 2021 (Boytchev 2021). As a consequence people began to see the vaccine as the least favourable option among the vaccines and began to refuse the jab—while there was no scientific evidence for its less effective nature. The actual problem, however, might have been data literacy among the stakeholders involved. The chaos in communication was followed by a legal dispute between the EU Commission and the manufacturer AstraZeneca over supply targets. The EU did not renew the order, as a consequence AstraZeneca was used less and less, especially in Germany.

The case of AstraZeneca shows that legislators and media have a special responsibility when it comes to communicating the selected candidates to the public. The careful consideration of procedures is not just a matter of regulatory decision-making based on scientific evidence under strict safety protocols, but also a question of public approval. Are the simplification and acceleration of procedures sufficiently communicated and legitimized? How can the shortening of clinical trials and, if necessary, the approval of a vaccine be sufficiently justified by the current situation in the context of the precautionary principle? Have medical and bioethical questions been considered at all stages? Who is currently able to produce vaccines and which suppliers are involved (bioeconomy and questions of distributive justice)? How can we ensure a continuous critical public discourse that reflects upon emergency use authorization? And who is responsible for initiating and maintaining said dialogue?

Communication and Resources
The focus of public debates seemed to be on aspects of timing. At the beginning the "race" between the candidates was at the centre of attention and not so much the scalability of the phenomenon itself—from the specific practical and ethical questions of synthetic vaccine development to the greater challenges for biotechnology. Future dialogue on these issues requires communication experts (especially for digital forms of exchange), resources and a willingness to experiment with different forms of participation—an open mindedness that should also reflect in funding structures.

Data Management
The current crisis also highlights questions of access and data security since most forms of activation would have taken place in the digital realm. Safety issues and data management will be one of the biggest concerns in designing and initiating future public dialogues. These aspects amplify with an issue of global importance—such as vaccination. Here, participatory elements in preparation of the potential testing, approval and marketing of a novel product require a heightened sensibility for cross-cultural communication and different data management/safety standards.

Training
The accelerated search for a vaccine also presents itself as a unique opportunity for education in the academic and public debate on wicked problems in the field of bioethics. If emerging biotechnologies aim to build a trustworthy framework, they should not miss the moment to engage with the public on the justifiability of current measures. However, this cannot be the task of the individual researcher or research groups and their partners alone. Rather, research councils and federal regulation institutions such as the PEI or the EMA should provide platforms of exchange and interaction to make sure that everyone involved, including representatives of different media outlets, are well aware of their own responsibility to shape the current public dialogue on synthetic vaccines.

References

Arvay C (2020) Genetische Impfstoffe gegen COVID-19: hoffnung oder Risiko? Schweizerische Ärztezeitung – Bull Méd Suisses Bollettino Dei Medici Svizzeri 101(27–28):862–864

Boytchev H (2021) Why did a German newspaper insist the oxford astrazeneca vaccine was inefficacious for older people—without evidence? BMJ 2021(372):n414. https://doi.org/10.1136/bmj.n414

Jiang S (2020) Don't rush to deploy COVID-19 vaccines and drugs. Nature 579:321

Kantar (2020) Zurückhaltung in der Öffentlichkeit gegenüber einem COVID-19-Impfstoff wächst – länderübergreifend. Retrieved from https://www.kantardeutschland.de/zurueckhaltung-gegenuber-covid-19-impfstoff

Lazarus JV et al (2020) A global survey of potential acceptance of a COVID-19 vaccine. Nat Med 27:225–228. https://doi.org/10.1038/s41591-020-1124-9

Lurie N et al (2020) Developing Covid-19 vaccines at pandemic speed. New Engl J Med 382(21):1969–1973. https://doi.org/10.1056/NEJMp2005630

Nguyen A, Catalan-Matamoros D (2020) Digital mis/disinformation and public engagment with health and science controversies: fresh perspectives from Covid-19. Media Commun 8(2):323–328

Paul KT et al (2022) Anticipating hopes, fears and expectations towards COVID-19 vaccines: a qualitative interview study in seven European countries. SSM Qual Res Health 2:100035. https://doi.org/10.1016/j.ssmqr.2021.100035

Shin MD et al (2020) COVID-19 vaccine development and a potential nanomaterial path forward. Nat Nanotechnol 15:646–655. https://doi.org/10.1038/s41565-020-0737-y

Siewert S, Dabrock P (2020) Pluralität achten – Nachdenklichkeit erzeugen – Orientierung anbieten: Der Deutsche Ethikrat als Spiegel und Katalysator gesellschaftlicher Debatten. Health Academy 3. In: Manzeschke A, Niederlag W (eds) Ethische Perspektiven auf Biomedizinische Technologie. De Gruyter, pp 246–261. https://doi.org/10.1515/9783110645767-023

Taylor A (2022) Why covax, the best hope for vaccinating the world, was doomed to fall short. The Washington Post. Retrieved from https://www.washingtonpost.com/world/2022/03/22/covax-problems-coronavirus-vaccines-next-pandemic/

Thanh Le T et al (2020) The COVID-19 vaccine development landscape. Nature 19:305–306

World Health Organization (WHO) (2022) COVID-19 Landscape of novel coronavirus candidate vaccine development worldwide. Retrieved from https://www.who.int/publications/m/item/draft-landscape-of-covid-19-candidate-vaccines

Zylka-Menhorn V, Grunert D (2020) Hoffnungsträger auch zum Schutz vor SARS-CoV-2. Dtsch Ärztebl 117(21):A1100–A1106

Chapter 7
Conclusion

Abstract In the conclusion, the authors argue that public participation should not be seen as a "one fits all" solution to a "crisis of trust" in emerging biotechnologies, but rather as a tool to work on collective visions for the future. To promote a productive approach in designing participatory strategies and formats, we need to consider the structural prerequisites, the value systems, and the communicative channels that enable participation in synthetic biology. The authors offer ten recommendations to foster a dialogue on how participation as an effective tool of governance can be pursued.

Keywords Recommendations Governance Synthetic Biology · Participation · Responsibility · Trust · Institutional Arrangements

Dynamic governance approaches need to go hand in hand with a more comprehensive perspective on public engagement: Participation is much more than public attitude research aiming to secure public acceptance. It holds the potential for securing a democratically legitimized innovation governance by being (i) an integral part of a precautionary approach, (ii) a decision-making tool for regulatory bodies, (iii) a crucial element of a trustworthy and transparent horizon scanning, (iv) a cornerstone for an authentic science communication.

This requires an awareness and willingness on the part of regulatory bodies and research institutions, researchers and social interest groups, producers and industry to develop holistic frameworks, where public engagement is no longer put on the sideline but becomes a valid operating mode of science monitoring and risk assessment.

The literature on synthetic biology considers public participation as an integral and potentially guiding element when it comes to the discussion of social and economic opportunities and challenges, ethical questions and even in the technological risk assessment of SB applications, but reports tend to gloss over the specific purposes, frameworks, prerequisites and modi of participation. Most calls for public engagement mean to tackle the grand (social and ethical) challenges. More and better public participation should be accompanied by more and better efforts in evaluation. Especially the impact of participation on SB governance (regulatory decision-making) still needs further research.

Discussions about responsibility seem rather one-sided, since they call upon researchers and research institutions to take up their role as guiding (and account-able?) council. This role, however, is poorly defined, its structural implications hardly considered. As trust is a prerequisite and at the same time goal of engagement, it becomes even more important to facilitate structural change and education that allows researchers to explore fundamental social, ethical and technological questions.

Scientists, however, cannot be expected to carry all the weight. According to the ETC Group (2012) researchers are not only responsible for establishing an open dialogue but are bound to a moral obligation that turns into an enforceable imperative, as they "must" report to the communities in which they work, to national govern-ments, and to the public via the internet. We would argue that most researchers are well aware of their responsibility but simply lack the training and financial and logis-tical resources to implement adequate participation events. One of the concluding questions of the "Dialogue on SB" illustrates this point: "How do we develop the capabilities for scientists to think through responsibilities?" (Bhattachary et al. 2010). What does it take for a researcher to act and identify not only as a scientist but also as a public engagement practitioner and science communicator? Should and can it be the responsibility of one person or council alone?

Here we would like to extend the notion of responsibility beyond one particular position and call to attention the multifaceted structural and cultural set-up necessary to incite effective engagement. In this vein, communication in its multifaceted and all-encompassing function is much underestimated. This also implies the question of who is involved and whose interests are represented.

There are ideas for institutional arrangements for public engagement, the ELSI program of the Human Genome project or various ethics commissions are just two examples. It seems, however, that the question of evaluation standards, of respon-sibilities, the broad individual awareness about the necessity of public engagement and the changes needed in research infrastructures to pursue a sustainable public dialogue is still missing. At the same time, we need to be aware of the national differ-ences in institutional contexts that shape the expectations and outcomes of public engagement exercises in each country. When regulatory and policy frameworks for governing emerging technologies are bound to European law, practitioners have to consider the spheres in which participatory activities actually make sense and will have a substantial effect on governance processes (Hansen 2010).

Public engagement should not be seen as a 'one fits all' solution to a (perceived and contested) "crisis of trust", but rather as a tool to work on collective visions for the future that should begin with a close inspection of the status quo, the reality of the lab, and the life of those who will be affected.

In the last section of this brief, we present 10 recommendations to foster a dialogue on how participation as an effective tool of governance in synthetic biology might be pursued.

References

Bhattachary D, Calitz, JP, Hunter A (2010) Synthetic biology dialogue report. Biotechnology and biological sciences research council and engineering and physical sciences research council. Retrieved from https://bbsrc.ukri.org/documents/1006-synthetic-biology-dialogue-pdf/

ETC Group, Friends of the Earth, CTA (2012) The principles of oversight for synthetic biology. ETC Group. Retrieved from https://www.etcgroup.org/sites/www.etcgroup.org/files/The%20Princip les%20for%20the%20Oversight%20of%20Synthetic%20Biology%20FINAL.pdf

Hansen J (2010) Biotechnology and public engagement in Europe. Palgrave Macmillan

Chapter 8
Recommendations

Abstract The authors provide ten recommendations for participation in synthetic biology: from questions of distributional justice to communicative efforts via resources and training to the technical infrastructure. Looking more closely at the structural prerequisites, value systems and educational needs of the stakeholders involved in SB governance allow us to acknowledge and deploy participation as a context-centered interest and purpose-driven process.

Keywords Trust · Justice · Translation · Time · Resources · Infrastructure · Communication · Evaluation · Training

As **trust** is a prerequisite and at the same time goal of engagement, it becomes even more important to facilitate structural change and education within research institutions that allow researchers and the public(s) to explore fundamental ethical, social, technological, and organizational questions of SB governance.

All stakeholders in SB should consider the distributional **justice** questions from the perspective of their particular space of enunciation. Issues of inequality require a structured dialogue that precedes and extends specific project frames and call for a broader awareness on all levels of governance.

Public participation should entail a systematic consideration of the voices and needs involved. A precise analysis of benefits and limits of participation requires us to ask: Who speaks and what are the fears/concerns/hopes expressed? Moving away from an abstract notion of the public(s) allows us to acknowledge and deploy participation as a ***context-centered interest and purpose-driven process***.

Public participation should be considered at all levels of governance: hard law, soft law, education, research impact, research infrastructure, depending on the purpose of engagement. In this respect, participation is not limited to the communication of research output but is seen as an ***end-to-end principle***.

An effective participatory approach requires **communicative efforts and translation** at all stages of governance. It needs communication experts to develop, plan, coordinate, implement and evaluate public exchange.

The **evaluation** of public participation activities in SB should be strengthened in order to determine what works well and what doesn't. Evaluation will also provide insights of the relative success of this kind of governance approach in SB.

Researchers need **time and resources** to experiment with different modes of participation and adequate forms of evaluation to secure the effectiveness of public-science interaction. Seed funding and small grants in preparation of bigger projects and research designs could provide the starting point for a well-balanced and transparent public-science dialogue and can be further used as training grounds for (aspiring) researchers to raise awareness for the different opportunities of a meaningful and trustworthy engagement.

Successful participation also depends on a clear and comprehensible **assignment of responsibilities**. The particular role of scientists, facilitators, research institutions/councils and participants needs to be well defined—especially when it is more difficult to mark the limits of commitment and involvement of certain public groups (as in the case of DIY-research).

Participation should enter the curricula of **SB training and study programmes** within a broader ethical and social context. Training should include the analysis of complex ethical questions and wicked problems of SB (and biotechnologies in general) *and* the practical implications of why and how to engage with the public(s). Collaborations with experts from different research fields and social contexts are appropriate (if not imperative) in view of the interdisciplinary nature of the field and the highly sensitive issues it addresses (and aims to solve).

Besides personnel, research projects need adequate resources for **technical infrastructure and supervision** when it comes to the implementation of digital participation platforms, an effective data management and improvements in data literacy (Fig. 8.1).

Fig. 8.1 Recommendations
for participation in synthetic
biology

PARTICIPATION IN SYNTHETIC BIOLOGY

10 Recommendations

1   TRUST

Facilitate structural change and education within
research institutions that allow researchers to
explore fundamental ethical questions, fostering
trust in science.

2   JUSTICE

Consider distributional justice questions from every
perspective involved. Issues of inequality require a
structured dialogue that precedes and extends
specific project frames and call for a broader
awareness on all levels of governance.

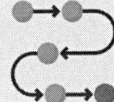

3   END-TO-END

Consider participation at all levels of governance:
hard law, soft law, education, research impact,
research infrastructure. It can be seen as an end-to-
end principle.

4   COMMUNICATION

Ask communication experts to help develop, plan,
coordinate, implement and evaluate participation at
all levels of governance. (Public) participation
requires tailored communicative strategies and
interdisciplinary translation.

5   INTEREST AND PURPOSE

Acknowledge and deploy participation as a context-
centered interest and purpose-driven process.
Public participation should entail a systematic
consideration of the voices and needs involved.
Who speaks and what are the fears/concerns/hopes
expressed?

Fig. 8.1 (continued)

PARTICIPATION IN SYNTHETIC BIOLOGY

10 Recommendations

6 EVALUATION

Strengthen the evaluation of public participation activities in SB. "Lessons learned" should be made accessible for future SB projects. A collectively maintained and updated platform for SB projects can facilitate the exchange of experiences on how to plan and install participatory elements.

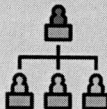

7 RESPONSIBILITY

Define the particular role of scientists, facilitators, research institutions / councils and participants. Successful participation depends on a clear and comprehensible assignment of responsibilities.

8 TRAINING

Make particpation an integral part of the curricula of SB training and study programmes. Training should include the analysis of complex ethical questions and wicked problems of SB and the practical implications of why and how to engage with the public(s).

9 RESOURCES

Consider the infrastructure required to provide authentic PE activities. Besides personnel, projects need adequate resources for technical infrastructure and supervision (including data security) to guarantee safe and effective digital participation.

10 EXPERIMENT

Allow for experimental and error-friendly participatory elements in funding guidelines. SB research infrastructures should enable scientists to implement authentic participatory elements in their project designs. This also entails time and resources to experiment with different modes of participation.